For Carla

Contents

Preface

Currently considered a bridge between basic and applied ecology, landscape ecology occupies an important new niche in ecology, representing a new star in the galaxy of the ecological sciences. However, the broad spectrum of conceptual and methodological approaches has created a non-focused science strongly influenced by the more dominant disciplines, such as landscape planning and restoration, forest management, landscape architecture etc.

The uncertain position of landscape ecology among the ecological disciplines is in contradiction with the general recognition that landscape is a spatial dimension in which important ecological processes occur, and landscape is becoming very popular in many ecology-related fields, from plant disease to animal behaviour.

There is a considerable amount of literature covering most of the themes of landscape ecology, and the theoretical frameworks are convincing enough to delineate new approaches and interpretations of ecological complexity. The literature is growing very fast, but as in many other pioneering disciplines a common framework is lacking and the topics often move from a human-oriented landscape ecology to simply large-scale ecology. In fact, landscape ecology is of two distinct kinds: the first is connected to European culture and its greater experience of landscape ecology, especially in the field of evaluation, management and restoration. The second has developed in North America over the last two decades, characterized by a theoretical basis and sophisticated methodologies producing complex studies beyond the ecosystem scale.

The presence of these two approaches is in some measure positive, and allows for the interchange of experience and points of view. In fact, in the last decade landscape ecology has found a multitude of cultural initiatives, centred on the International Association for Landscape Ecology (IALE), and a number of working groups have been successfully established in local ecological societies and in other NGO organizations.

Two possibilities exist to expand landscape ecology: one consists of developing new research, and the other in developing a good educational framework. Both are important and not in conflict. In this spirit I have prepared this book, with the aim of summarizing the best theories, concepts, principles and methods in landscape ecology. It is an attempt to reinforce the ecological research perspective, to consolidate principles and methods, validate procedures and reconcile different positions, including the geobotanic, animal and human perspectives.

The concept is very simple. I have no ambition to present new ideas and theories: I have worked to create a tool mainly for classroom use but also appealing to a broad range of scientists and practitioners dealing with landscape ecology and its problems.

The selection of a simple and sufficiently contemporary objective conceptual path is difficult because of the personal interpretation of the discipline. The theoretical basis, the contribution of other disciplines, emerging processes and patterns, managing applications and methods are the main steps that I have utilized in this exciting journey.

Each chapter contains a summary of the main points discussed and suggested reading. The book is not comprehensive, neither for topics nor for references, but I have tried to maintain a good balance between the types of literature suggested. Often my background as a naturalist has prevailed, but the percentages of literature for the different topics have been respected and it is hardly surprising that animal studies dominate in landscape ecology. In some cases it was not easy to make the best choice, either because of the great number of studies (e.g. on the effect of fragmentation on animal populations) or because too few were available (e.g. soil landscape and flux of nutrients).

Some comments are necessary to explain the general outline of the book. In the introduction I have tried to avoid long historical perspectives, focusing more on the real object of landscape ecology and on good definitions. I have also highlighted the contribution of related ecological disciplines in the creation of a strong conceptual framework.

The description of new theories, such as percolation, metapopulation, hierarchy etc., precedes the more detailed approach. Emerging processes (fragmentation and disturbance, connectivity and ecological fluxes) and patterns (heterogeneity and ecotones) occupy the central part of the book. Landscape dynamics, management and nature conservation are all extensively described.

The last chapter is devoted to methods. Special emphasis is placed on indices describing the structure of the landscape mosaic, from Euclidean to fractal geometry. Geographical information and global positioning systems have been included as indispensable tools. Remote sensing procedures and spatially explicit models occupy the final part of the book, with very simple routines for measuring landscape structure and complexity. These routines may be improved and incorporated into more sophisticated programs. I have tried to encourage

measurement of the landscape using simple tools, because of the frustration felt by those who do not have access to the expensive and powerful computation and remote sensing facilities of super-specialized advanced research centres.

I am perfectly aware of the limitations of this book; I am also conscious that many perspectives have not been discussed, such as the socioeconomic implications.

Most of the pictures and examples are from my preferred study area (northern Apennines, Italy). The environmental and cultural complexity of this region, like most of the Mediterranean basin, is an exciting field in which to test and apply landscape principles and methodologies, and is an inexhaustible source of scientific creativity.

I am indebted to many people, and in particular to Zev Naveh for his invaluable encouragement. I am grateful also to Francesco Di Castri for his friendship and support during the preparation of this book.

Foreword

Landscape ecology has its roots in the long tradition of central and eastern European geobotanists, ecologists, geographers, landscape planners and architects who were not content with the present state of their sciences and professions. They strived to present their rich and heterogeneous landscapes in more holistic ways, as the spatial and functional integration of nature, humans and land, so that their studies could be of practical value in landscape appraisal, planning, management, conservation and restoration. However, chiefly because of language and cultural barriers it remained a rather restricted 'continental' science until it was joined more than twenty years ago by the 'second generation' of a large group of far-sighted – and chiefly North American – ecologists and geographers. These realized the theoretical and methodological relevance of landscape ecology and the need for broadening the spatial scales of ecosystem ecology for the study of the ordered complexity of natural and cultural landscapes. The two groups joined together and founded the International Association of Landscape Ecology (IALE). Fortunately these developments coincided with the dramatic advances in remote sensing and satellite imaging with finer and finer resolutions over larger and larger areas and with the progress in processing larger masses of data in smaller and cheaper computers with more sophisticated and comprehensive modelling methods. Since then landscape ecology has spread its wings all over the world both in industrialized and developing countries as one of the youngest and most dynamic branches of contemporary environmental science.

The author of this book, Dr Almo Farina, is the first one of the 'third generation' who not only followed in the footsteps of both these founder groups but contributed a new milestone to its further development and especially to the education of the next generation of landscape ecologists, academians and professionals. He took upon himself the challenge to provide a meaningful synthesis of what he considered to be the 'best theory, concept, principles

and methods' which are presently applied in a multitude of landscape-ecological studies and are published in the journal of 'Landscape Ecology' and in many other journals and scientific publications. Presenting in a lucid way some of the most relevant new ideas, theories and paradigms, he succeeds also in reconciling the diverse geological, biological and human perspectives. At the same time he provides his own original, well-versed and well-balanced contribution to contemporary landscape ecology as a holistic, quantitative and problem-solving oriented science for the promotion of sustainable, healthy and productive landscapes.

Although dealing in a systematic way with a large body of rather complex scientific information, such as fractal dimensions, numerical and spatial data processing and geographic information systems, this book is far from being dry, technical and detached from reality. On the contrary it is very lively with many fine illustrations and with many practical examples. While reading through its chapters I could sense that it was written by one who is eager to communicate not only his own knowledge and holistic perception of landscapes as a hybrid nature-culture gestalt systems, but also his close personal attachment to the biological and cultural assets of his Appenine mountain and rural landscapes in which he grew up, and lives and works, and where he carries out his own research.

A great advantage, in my opinion, is the fact that this book was not written by a purely academic scientist, spending most of his time sitting in an office behind a computer, trying to publish as many possible 'scholarly' works to further his own reputation.

Dr Farina started his professional career as a high school teacher in biology and is still very active in public education as Director of the Museum for Natural History at Aulla, Italy (which was established and is maintained thanks to his initiative to preserve one of the most outstanding historical landscape monuments of this region). He started his research as an enthusiastic ornithologist but very soon realized the great potentials of landscape ecology, which fitted very well with his deeply ingrained perception of the landscape as a whole, and his intellectual abilities for acquiring the most advanced methods available and to turn these into practical tools for the study, management and conservation of landscapes. Dr Farina is not only active in these local issues but is also deeply involved in the broader issues of the future of Mediterranean landscapes in Italy and in the Mediterranean Basin. He also served for four years as the secretary of the IALE.

I am confident that this book will serve very well as both a textbook and as a handbook for those involved in landscape ecological study, research and education, as well as for many others from closely related fields of natural and human sciences dealing with land use. I am also hopeful that it will help to bridge the gaps between these different fields so that landscape ecology can be realized as one of the most important integrative environmental sciences in this crucial transition period from the industrial to the information age.

Zev Naveh
Haifa, Israel

Introduction to landscape ecology

1.1 INTRODUCTION

Landscape ecology is one of the youngest branches of ecology. It evolved after World War II in the countries of central and eastern Europe and only recently expanded into a unique, dynamic and integrated global science. Its roots are deep in geography as well as in geobotany and land management.

The German geographer and scholar Alexander von Humboldt, 200 years ago, regarded the landscape as 'the total character of a region', but the term landscape ecology was coined by the German biogeographer Carl Troll at the end of the 1930s. Troll hoped that a new science could be developed that would combine the spatial, 'horizontal' approach of geographers with the functional, 'vertical' approach of ecologists.

Landscape ecology was born as a human-related science (Naveh and Lieberman 1984) but it has recently been accepted that the landscape is very promising for ecological studies (Risser *et al.* 1984, Forman and Godron 1986, Turner 1989, Farina 1993a, Forman 1995). In 1986 its principles and methods were introduced to a large audience of ecologists at the IVth International Congress of Ecology in Syracuse, New York.

The birth and development of landscape ecology have been a progressive, dynamic and global process

– still active – covering many fields of ecology and related sciences, such as geography, botany, zoology, animal behaviour and landscape architecture. The landscape perspective is full of promise for the realization of the integration of different sciences.

The scale of the landscape comprises a complete set of socioeconomic and ecological processes. All combine to form the real world, but considered separately they acquire a fictional character. The change of approach from the study of distinct ecosystems to that of the landscape probably depends on the core of unresolved questions and shaded areas that the ecosystem approach has maintained.

There are problems with the way in which these various disciplines will interact, but space is recognized as a new frontier of ecology and the landscape is one of the main components of this space.

1.2 CONTRIBUTION OF DIFFERENT DISCIPLINES

Recognition of the landscape as a suitable spatial scale at which to investigate ecological processes has been a complicated process rooted in apparently distant theories. It is our opinion that the island theory (MacArthur and Wilson 1967) and the focus on ecological geography (MacArthur 1972) are two fundamentals that have opened the way to the development of modern landscape ecology. That it is an important component in determining the diversity of life forms, and that most the ecological patterns and processes have unique shaping factors, may be considered the most influential paradigms to introduce space as a fundamental element of ecology *per se*.

The presentation of ecological systems as components of a nested hierarchy (Allen and Starr 1982, O'Neill *et al.* 1986) has strongly contributed to link different paradigms and theories incorporating the concept of scale. New concepts such as fractal geometry have been introduced in the ecological field (Mandelbrot 1975) to investigate the complexity of nature. This complexity has for decades discouraged ecologists from taking into consideration some of the more important interfaced ecosystems, such as coasts and marshes.

The new ideas about heterogeneity (Kolasa and Pickett 1991) and the role of the disturbance regime (Pickett and White 1985) in the ecological processes represent a further stage on which new paradigms such as ecotones (Hansen and di Castri 1992), related processes like connectivity (Merriam 1984) and theories such as metapopulation (Gilpin and Hanski 1991) have been implanted. As a consequence of the acknowledged heterogeneity of the landscape the source–sink paradigm developed by Pulliam (1988) recognizes and assigns new roles to the patches composing the landscape mosaic.

The concept of landscape scale frequently appears in the scientific literature, ranging from the soil science (Buol *et al.* 1989) to the new perspectives in geoecology (Huggett 1995).

1.3 DEFINITIONS OF LANDSCAPE

Landscape ecology is a young science without unique definition and concepts. A wide range of disciplines converge in the direction of landscape ecology; for this reason there are several definitions of landscape:

- 'the total character of a region' (von Humboldt);
- 'landscapes dealt with in their totality as physical, ecological and geographical entities, integrating all natural and human ('caused') patterns and processes...' (Naveh 1987);
- 'landscape as a heterogeneous land area composed of a cluster of interacting ecosystems that

is repeated in similar form throughout' (Forman and Godron 1986);
- a particular configuration of topography, vegetation cover, land use and settlement pattern which delimits some coherence of natural and cultural processes and activities' (Green *et al.* 1996);
- Haber has defined the landscape as 'a piece of land which we perceive comprehensively around us, without looking closely at single components, and which looks familiar to us' (pers. com. 1996).

This last definition is more general and more suitable for defining the landscape as perceived by all other organisms, from plants to animals. This allows a promising new field of research and speculation to open, on the importance of spatial arrangement of patterns and processes for the functioning of organisms, groups and ecosystems. In this way it is possible to unify different views and concepts related to the landscape by scalar normalization.

These different definitions have been produced by different cultural and scientific approaches. Most actual landscape ecology deals with human-disturbed ecosystems, which is inevitable because of the widespread distribution of human populations across the earth. However, pristine areas could also be efficiently approached using landscape ecology.

Generally the landscape is considered a broad portion of a territory, homogeneous for some characters such that it is possible to distinguish the type by the relationships between structural and functional elements. The broad scale of a landscape implies that many processes can be observed in the interior across a broad spectrum of temporal scales (Fig. 1.1).

Landscape ecology is a study of complicated systems but needs to be referenced to an organism to be better understood (Turner *et al.* 1995). For example, the landscape as perceived by humans is different in size from that perceived by a beetle (Wiens and Milne 1989). Hence to humans the landscape is a broad-scale area composed of a mosaic of patches or ecotopes, into which we introduce physical, biological and cultural elements. In dealing with the beetle's landscape we necessarily reduce the physical and biological entity to the beetle's assumed perception of that landscape.

Landscape ecology offers an extraordinary opportunity to carry out new experiments in which the contribution of different disciplines is essential,

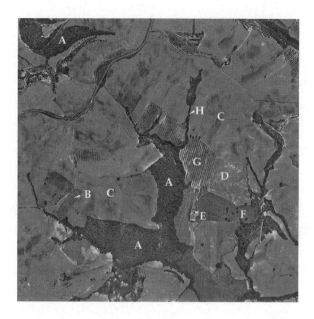

Figure 1.1 Aerial view of a rural landscape of southern Tuscany. The area is a mosaic of woodlot (A), olive orchards (B), plight fields (C), set-aside fields (D), farms (E), scattered trees (oaks) (F), vinyards (G) and strip vegetation along streams (H). This human socio-economic dimension is the basis for any land management and political decisions. Landscape ecology functions as a bridge between ecology and human-oriented disciplines.

so distinct landscape ecologies are now available, from the more sophisticated in which human perception is compared with natural processes, to the more simple in which the ecology is approached using an enlarged spatial scale.

1.4 THREE VIEWS OF LANDSCAPE ECOLOGY

Becaue of its different and complicated roots landscape ecology may be approached in several ways. Relevant books have emphasized this fact in their introductory remarks (Naveh and Lieberman 1984, Forman and Godron 1986, Forman 1995, Zonneveld 1995). However, these different approaches need to be reconciled. Historically this discipline originated at the human-perceived level

and the first descriptions of families of processes were strictly linked to human life. As landscape ecology developed into the study of the spatial arrangement of patterns and processes concerning the soil, vegetation, animals and humanity, especially in North America, a formidable tool has been launched.

At least three perspectives in landscape ecology can be distinguished:

- Human: In the human perspective landscape is grouped into functional entities that have a meaning for human life (Fig. 1.2).
- Geobotanical: The spatial distribution of the abiotic and biotic components of the environment, from the soil landscape to that 'perceived' by plants, and to the distribution of plant entities as communities, woodland, prairies, woodlot etc. For plants perception must be considered in a broad sense as the range of sensitivities of plant life requirements, and their capacity to incorporate information from their environment. This is directly connected to adaptation, colonization and survival in the face of natural and human-related stress (Fig. 1.3).
- Animal: This last perspective is conceptually related to the human-perceived landscape, although there is a substantial difference in that the approach is on a direct or predictable animal species-specific scale (Fig. 1.4).

Each of these three approaches explores a range of patterns and processes which in the last analysis are components of the whole biological and ecological system.

There are more commonalities than differences in the three approaches, such as space and the spatial arrangement of processes and patterns. And although it is not always obvious from studies and research in landscape ecology, landscape ecologists are really aware of the interactions between the human world and nature, and also their interdependence. The requirement to place a pattern or a process in space at the correct scale is a common goal of landscape ecologists.

The human dimension of landscape is probably the most complex approach because of the overlap and the interweaving of animal and cultural components of humanity. This dimension is related to processes that have a broad temporal and spatial scale. The biological dimension of humans may be

(b)

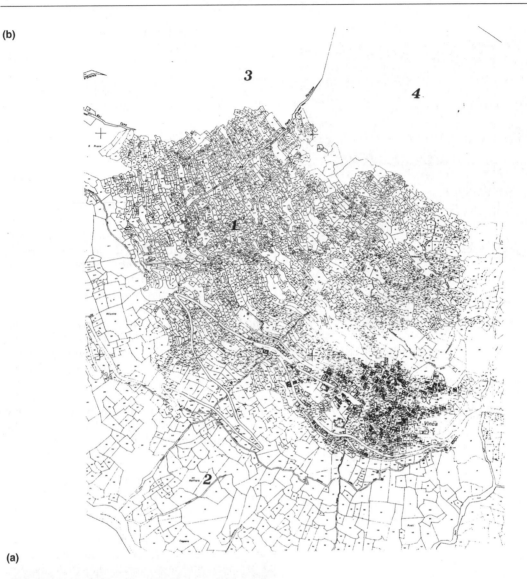

(a)

Figure 1.2 Human (rural) landscape. (**a**) The Vinca village (northern Apennines) and the neighbouring landscape composed of (1) terraced fields, (2) sweet chestnut orchards, (3) calcareous bare soil and pastures and (4) pine plantations. This system has deep roots in the Vinca community, affecting its socioeconomic structure. Comparing this land cover with the cadastral map (**b**) it is easy to recognize small properties (a few square metres) in the terraced area (1), medium-sized properties in the sweet chestnut orchard (2), and no property division in the bare soil, pastures and plantations (3 and 4). In this last case low-quality soil is considered an undivided village property that all inhabitants can use for grazing and logging.

(a)

(b)

Figure 1.3 Mountain ridge landscape (geobotanic) (M. Prado, 2000 m, northern Apennines). (**a**) Rock debris, bare soil and vegetation patches create a complicated mosaic in which climatic constraints (snow cover, wind) and sheep grazing shape the vegetation. (**b**) At small scale (1 m) vegetation patches (*Trifolium alpinum*; *Alchemilla alpina*) are shaped by interspecific competition, microclimate and soil attributes.

compared with the animal dimension, but the cultural component of humanity is unique and this is the one that interacts most with the environment, especially because of the dominance of human culture and the ability, using technology, to overcome natural and ecological constraints and limitations. From these three perspectives the challenge to com-

bine theories, paradigms and models produced by the traditional monodisciplinary approaches is evident and a recurring theme.

The ability of landscape ecology to serve as a framework for an advanced comprehension of ecological processes can be realized only if we recognize these different approaches, which together pro-

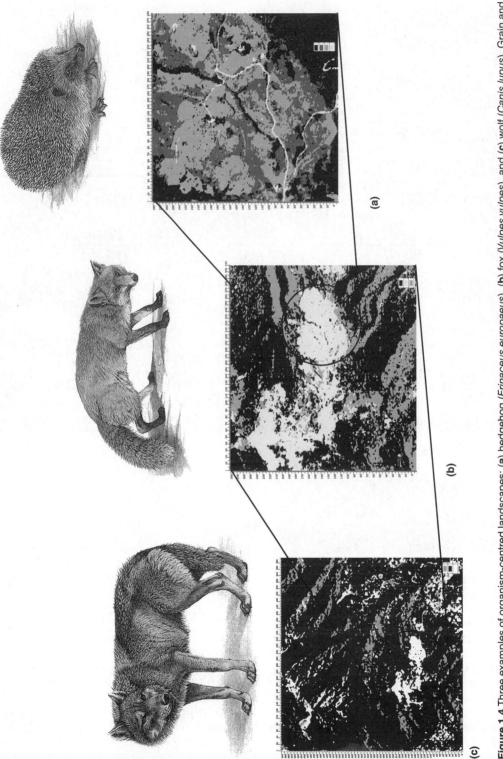

Figure 1.4 Three examples of organism-centred landscapes: **(a)** hedgehog (*Erinaceus europaeus*), **(b)** fox (*Vulpes vulpes*), and **(c)** wolf (*Canis lupus*). Grain and extent appear species-specific. The images were drawn from aerial photographs (1988–89) of the Monte Orsaro range, northern Apennines, and classified according to the three major land covers: open spaces, shrubs and woodlands. The hedgehog landscape is represented by prairies and shrubs bordered by woods (the size of area is approximately 300×300 m). The fox landscape is represented by prairies, fields and dense and open woodlands (the area is approximately 3×3 km). The wolf landscape embraces a catena range and is represented by prairies and pastures, woodlands of different types, clearings and hedges (the size of the living area is more than approximately 10×10 km).

duce the best information available. The complexity of the environment is so great that even this approach can lead to only a partial understanding. Landscape ecology cannot explain all the processes, but can undoubtedly help us to understand the complexity, i.e. the interrelationships between different processes and patterns. The common denominator is the spatial dimension of the processes, and their importance to other spatial and non-spatial processes and related aspects is one of the relevant foci of landscape ecology.

The risk of considering landscape ecology from an exclusively anthropocentric viewpoint is very great, and could produce a dogmatic discipline lacking theoretical and experimental verification. On the other hand, simplification on a broad scale is an unacceptable and reductionist view. The main strength of landscape ecology consists in its ability to transfer information across different families of processes occurring at different spatial and temporal scales. The spatial dimension has been recognized as extremely important in the progression from a topological approach to one in which the real world is studied.

There exist a human landscape, a plant landscape and an animal landscape, all composing the environment of the planet and the context in which ecological processes act. This premise is extremely important as a basis for understanding the choice of arguments and their position in this book. Landscape ecology may be considered as a new science, or as a pioneering approach of ecology if geobotanical and animal perspectives are taken into account. In the first case the study of human-related processes accompanied by processes created by other entities (plants, microbes, animals) creates a more sophisticated context greatly appreciated by scientists working in disciplines like sociology and architecture.

The human and animal perspectives have some points in common when the scale of animal perception is close to the human one. In this case all benefits gained by humans in shaping and/or controlling the landscape are shared by some species of animal. Landscape designs carried out by experienced architects can change the availability of resources in such a way as to favour certain mammals and birds, for instance. Figure 1.5 shows an example of how to create more complexity and edges in a restored rural landscape in the UK (Forestry Commission 1991). Many highly scenic landscapes are attractive to tourists as well as wildlife.

A note of caution should be introduced regarding an approach that ignores ecological criteria and models. In fact, in this case the benefits are shared between humans and those animals that are adapted to live together with humans.

However, when changes are requested in a pristine environment such as an old-growth forest in the Pacific USA, this approach is both insufficient and dangerous. For some species, such as large carnivores or nocturnal raptors, or most of the neotropical birds that use the landscape in a very complicated way, more information is needed and a landscape plan that, at first sight, is a nonsense from the human point of view. In such a case we need a strong ecological background and good models working in an explicit way, considering the physical and biological components.

There is also another limit to this approach in that, without a knowledge of population dynamics spanning the lifecycle of a species, we could create sink and not source habitat landscapes, thereby accelerating a decline in the target organisms. To improve this anthropocentric approach a strong interaction is necessary between the geobotanical and animal perspectives that can be incorporated into common practice.

Predicting change could be extremely important for the survival of many species and entire ecosystems. Figure 1.6 illustrates the scenario of private versus public cutting of forest landscape in Oregon between 1972 and 1988 (Spies *et al.* 1994): private cutting has caused a dramatic decline in pine forests owing to unselected cutting; the public forests have been fragmented but the pine matrix survives.

Finally, landscape ecology is one of the most promising ecology-related disciplines, which are highly differentiated but have a common core based on the finite dimension of the studies, overlapping of data and information on the real world, and on the new 'virtual' processes linked to information transfer at a global scale.

1.5 SPACING – THE PERCEPTION OF THE LANDSCAPE

Space may be considered as 'The final frontier for ecological theory' (Kareiva 1994). We refer to spacing as the reaction of an organism to its perception of its environment (the landscape).

Spacing, or spatial arrangement, is a scaled property of living organisms, from individuals to

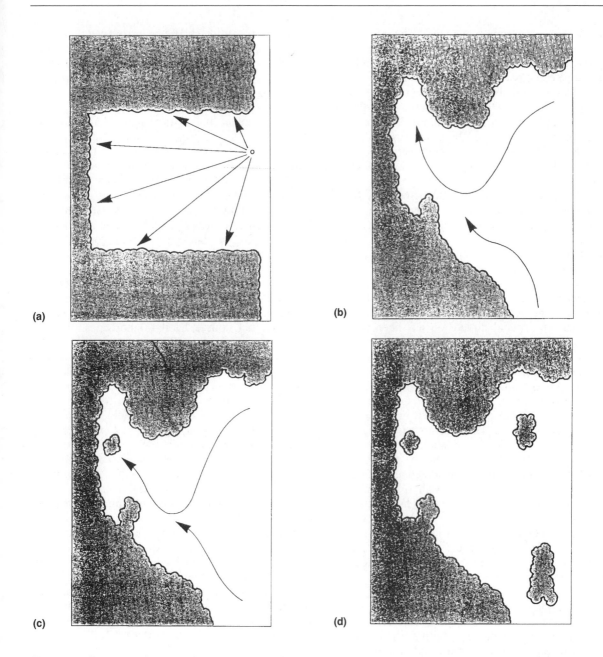

Figure 1.5 Structuring the landscape consists not only in changing the scenery value but also creating or destroying structures with spatial and ecological attributes. In this case the activity was devoted to increasing the spatial complexity of a clearing from (a) to (d) (UK, Forestry Commission 1991).

populations, communities and metacommunities. Organisms react to external stimuli because of their biological need to optimize resources and the energy to provide such resources. Spacing is the ecological reply of an organism to non-uniform distribution of resources (habitat suitability) and to inter- and intraspecies competition in space and time. This concept is central to landscape ecology.

Private   Public

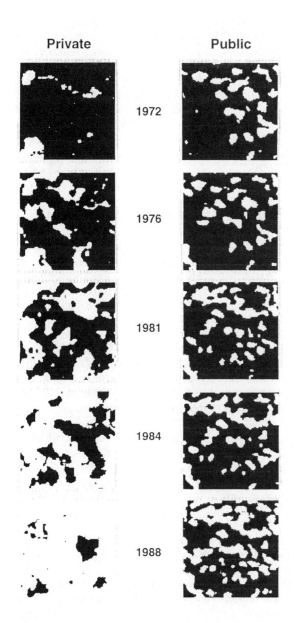

1972

1976

1981

1984

1988

Figure 1.6 In Oregon forested landscape changes according to private and public ownership type. Black indicates conifer type, white other types of woodland (from Spies *et al.* 1994, with permission).

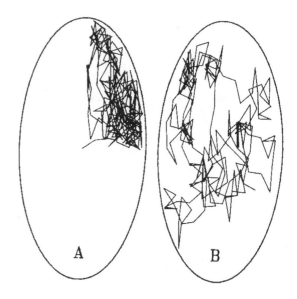

Figure 1.7 Different patterns of home range used by mammals. The path in **A** is more concentrated than that in **B**, although the home range is the same (from Swihart *et al.* 1988, with permission).

Grain is the minimum area at which an organism perceives and responds to the patch structure of landscape (Kotliar and Wiens 1990) (Fig. 1.7).

Extent is the coarsest scale of spatial heterogeneity at which organisms react.

The response of a biological entity to different availabilities of patch habitat will be considered across different spatial and temporal scales. Pearson (1991), conducting experiments on the foraging patterns of field sparrows and white-throated sparrows, described the different spatial arrangement of individuals foraging in a cage: field sparrows fed close to each other, but white-throated sparrows used the entire available area to avoid violating individual spaces (Fig. 1.8).

This behaviour can be better appreciated in a population of frugivorous birds in winter. During the warmer days flocks are relaxed and species dispersed in the environment, but when the temperature drops the interindividual distance is immediately reduced and species can coexist close to each other where resources are still available. So, spacing is determined not only by external cues and habitat availability but also by physiological constraints, and this creates a more complex scenario.

Spacing depends mainly on resource availability. Plants react to resource availability by arranging themselves in a finite and predictable pattern. The African acacia savanna is a typical example. In

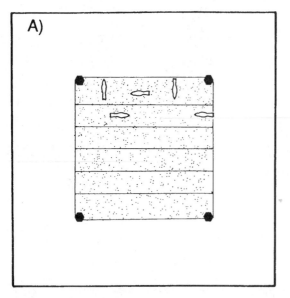

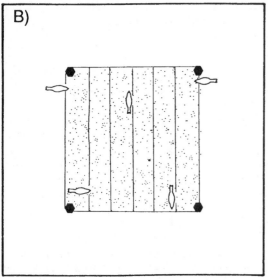

Figure 1.8 (A) Field sparrow individuals can forage close each to other, whereas **(B)** white-throated sparrows maintain a high interindividual distance during food searching. From an experiment by Pearson (from Pearson 1991, with permission).

Europe this patterning is represented by the Spanish *dehesa* and by the Portuguese *montado* (Fig. 1.9).

Animals are very sensitive to their surroundings, especially as regards antipredatory behaviour, intraspecific competition etc. There are species that forage separately and show a strong territorial behaviour. One example is the European robin (*Erithacus rubecula*), which defends an area and fights any individual that does not respect its home range. The defended patch in many cases is composed of a mosaic of fields, hedgerows and woodlots of different extents according to the territory patch quality, but also according to the competition, the abundance of other individuals, and finally by the attractiveness of the area generally (Farina 1993b).

1.6 BEHAVIOURAL ECOLOGY AND LANDSCAPE ECOLOGY

Among the topics considered by behavioral ecology, such as the animal decision making for movement, dispersal and habitat selection, which determines their dynamics and spatial distribution, many are also common to landscape ecology. Unfortunately, there are at present too few contacts between the two research fields for the problems to find a common interface (Lima and Zollner 1996).

Perceptual range is defined as the distance from which a particular landscape element can be perceived. This represents the species-specific window of identification of the greater landscape, and plays a fundamental part in the survival of individuals. A species that has a low perceptual range can expect a high risk of mortality, spending more time searching for an available habitat than a species with a high perceptual range. In the case of habitat heterogeneity and low perceptual range, a species has a high risk of predation.

The perception of landscape patchiness is fundamental for animals, but little information can be found in the literature (Johnson *et al.* 1992). For instance, the foraging activity of bighorn sheep (*Ovis canadensis*) in a heterogeneous landscape has been explained by Gross *et al.* (1995), applying a nearest-neighbour rule. Some species, such as birds, have a great perceptual range, especially in certain seasons. Chemical signals may improve this capacity in many arthropods, and honey bees have a better sense of orientation when the landscape is rich in topographical irregularities (Southwick and Buchmann 1995).

Patch selection and movement are also largely dependent on conspecific attraction: an individual tends to settle in a patch occupied by another conspecific (see Smith and Peacock 1990, for a review). At the landscape level animals tend to con-

Figure 1.9 The *dehesa* is a park-like landscape in which plant spacing is determined by water availability. In this savanna-like landscape, created mainly by humans, wild and domestic grazers play a fundamental role in storage of the available biomass resources.

centrate in certain patches, and dispersal is a deterministic process. This is well documented in birds outside the breeding season, during either foraging or roosting behaviour.

The perception of corridors is another focal point. Some species extensively use corridors that have a recognizable structure, such as hedgerows, but in many cases corridors are perceived by animals through a species-specific integration of visual, acoustic and olfactory cues.

By using spatially explicit models and the source–sink principle we can assume that animals select the most suitable patches from among the ones available in the landscape.

Animal movement and patch selection are determined by many internal and external cues, using a spatial memory, cognitive maps and conspecific attraction. A new and unexplored opportunity to understand the perception capacity of animals is

created by the human-altered landscapes in which animals have to face new landscape patterns, which in many cases are truly novel and different from the ones in which they evolved. For instance, forest animals now live in a more fragmented landscape and their maladaptation (Blondel *et al.* 1992) to the new patterns can be used as tool to investigate the evolutionary process.

1.7 STRUCTURING THE LANDSCAPE

Considering that a landscape is intrinsically heterogeneous, from human to beetle scale, the components of this mosaic are represented by individual patches inserted in a matrix, by which we mean the dominant cover (Fig. 1.10). This role applies at all scales and with all approaches.

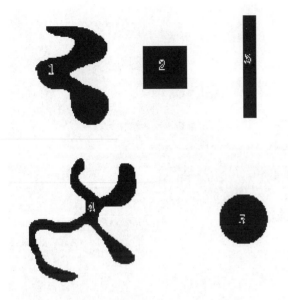

Figure 1.10 Patch size and shape are important attributes influencing abiotic and biotic fluxes. Different shapes, both regular and irregular, can be commonly found in nature. The more irregular a patch the more edges are available. This has tremendous implications for plant dispersion and animal movement. Using the ratio L/2√Aπ, where L = patch perimeter and A the area, it is possible to evaluate the amount of edges. The circle has a ratio of 1 and represents the euclidean figure with the minimum ratio between perimeter and area (see Chapter 8).

Landscape can be considered a new, higher level for biological complexity (Lidicker 1995), and is extremely useful for understanding complex processes. A landscape is composed of patches of more than one community type. The spatial arrangement of patches, their different quality, the juxtaposition and the proportion of different habitat types are elements that influence and modify the behaviour of species, populations and communities (Figs 1.11, 1.12).

1.8 PRINCIPLES OF LANDSCAPE CLASSIFICATION

It is possible to classify a landscape and its component patches using a number of approaches that can be either anthropocentric or dependent on perceptive capacity.

- Structural patch: Generally composed of a soil type overlapped by associations of vegetation;
- Functional patch: An area homogeneous for a function or a physical descriptor, such as altitude, temperature, moisture, light penetration. In this category we can include the ecotope, a selection of characteristics which, when met together, determine a unique character at a higher level. Ecotope classification is subjective and directed at a goal. Often it represents an attempt to find a group of spatially coincident characters to correlate with the distribution of a species, a behaviour or, more generally, a process.
- Resource patch: Mostly related to animal ecology, a landscape can be described a combination of resource patches. These are considered part of an animal's home range in which food or nesting sites or roosts are easily available, i.e. part of the home range in which specific functions are concentrated. They have an effect on individuals and are considered equal to or smaller than an individual home range (Fig. 1.13).
- Habitat patch: May be defined as distinct plant community types that are generally larger than an individual home range (Ostfeld 1992). Different groups of organisms can share the same habitat patch (Fig. 1.14).
- Corridor patch: Although the definition of a corridor and its use is controversial, we consider a corridor patch to be a portion of the land mosaic that is used by an organism to move, explore, disperse and migrate. Often the corridor concept is associated with a narrow strip of land, but for more details see Chapter 4. Generally we associate corridors with a special feature of an organism that is accomplished outside its normal life (Table 1.1).

Classification is a relevant procedure in the study of the land mosaic, especially as regards the human perspective. This approach is generally used by landscape ecologists interested in studying the inter-

Table 1.1 Numeric values of perimeter, area and edge amounts of differently shaped patches of Fig. 1.10

Patch	Perimeter	Area	L/2√Aπ
1	659	10027	1.857
2	277	4900	1.119
3	373	3652	1.745
4	1125	9736	3.217
5	269	5222	1.051

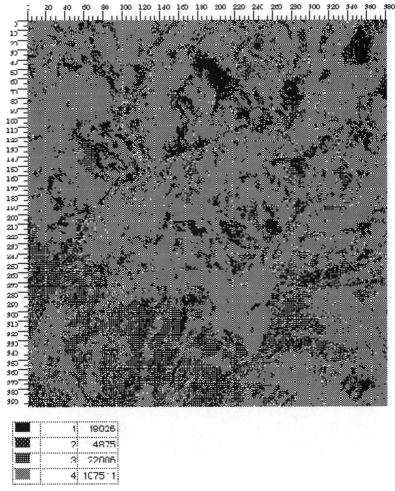

Figure 1.11 Representation of an 8×8 km area in the Apennines (data source Landsat 1990, Magra river watershed, northern Apennines, Italy), classified according to three categories of land cover: 1) woodland, 2) *coltura mista* and 3) prairie. The woodland represents the matrix of this portion of the landscape owing to its dominant extension; cultivation and prairie are considered patches in such a matrix.

action between human activity and the landscape. It is particularly useful for the preparation of master plans, in planning nature reserves and in general as

Table 1.2 Number of patches, min, max and mean values for perimeter and area of subset image of Fig. 1.14

	Perimeter	Area
No patches 138		
min	2.83	1
max	305	705
mean	20	31

a guide to many types of land management. There are no precise rules but these change according to the purpose, the scale of investigation, the time and the financial resources available (Table 1.2).

In order to produce a good and useful classification large amounts of information are necessary. The main sources are aerial photographs, satellite digital images, cadastral maps, geological, hydrological and soil maps, geographic and biothematic maps (vegetation, land use, animal distribution) (Fig. 1.15).

All this material should be harmonized and then used to produce different types of maps. The first step is to create physiotopic maps.

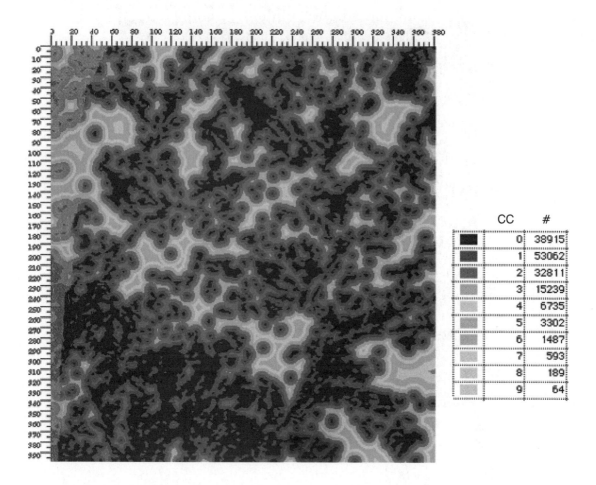

CC	#
0	38915
1	53062
2	32811
3	15239
4	6735
5	3302
6	1487
7	593
8	189
9	64

Figure 1.12 Connectedness between patches (see Chapter 4) is an important attribute of a land mosaic. In this case the connectedness has been calculated as distance (nine categories) from prairie and cultivation patches (CC). (# = number of cells for each CC category.) Detail from Fig 1.11.

A physiotope may be defined as a spatial unit characterized by relatively homogeneous abiotic state factors. Generally a physiotope is classified using geology, aspect and slope rate. It is the basis for further landscape classification (Vos and Stortelder 1992) (Fig. 1.16).

In soil classification the elementary unit is the pedon; a polypedon is considered to be a grouping of contiguous pedons. The boundaries between different polypedons may be sharp or gradual. The physiotope may be considered to be a pedon plus other edaphic and microclimatic characters.

Adding to the physiotope vegetation, land use and humus forms, it is possible to classify the ecotopes. These discrete units are further distinctions of the land type that can be found in a region. There are no precise rules about ecotope classification, but approaches vary according to the different purposes and details requested. The classification of elementary landscape units or ecotopes may be compared to the site concept, or 'facies' (Woodmansee 1990), which are defined as a biotic community existing in a soil polypedon.

The characteristic spatial arrangement of ecotopes creates a landscape, but when some ecotopes are found more associated than others it is possible to distinguish land units. A cluster of sites is the highest level of organization that can be compared, from microchore to mesochore (Zonneveld 1995). The landscape is composed of site clusters, the

(a)

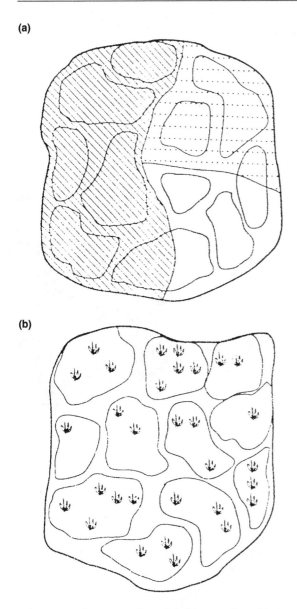

(b)

Figure 1.13 Representation of habitat patches **(a)** overlapped with small-mammal territories. Three habitat patches are represented. Generally territories belong to a unique habitat patch. **(b)** Resource patchiness represent a further division of a habitat patch (from Ostfeld 1992, with permission).

highest level being represented by land systems, regions, ecoregions, climatic zones etc.

We can consider this classification as hierarchical. At the lowest level we find the physiotope (although in many cases the physiotope can be larger than the ecotope), then the ecotope, the land unit and the land system.

The ecotope represents the topological dimension of a landscape, whereas the land unit and higher aggregations form the chorological dimension. The ecotope concept can be used with an anthropocentric perspective and bounded according our perceptual capacity. In reality it is applied more generally to a classification of the landscape, comprising plant and animal landscapes. Although many classifications are proposed (as reported by Naveh and Lieberman 1984, Woodmansee 1990, Zonneveld 1995), to be efficient and rigorous classification should mention the scale of the hierarchy adopted and the reasons for its choice.

In many human-modified landscapes the spatial arrangement of vegetation and land use is so bounded by human activity that the land mosaic appears very patchy, and the contrast between patches so high (e.g. cornfield bordered by old-growth forest) that it is difficult to imagine any other spatial arrangement of the patches. Plants, animals and microbes are forced to live in such contrasting land types.

Often we have insufficient information about the ecological validity of our ad hoc classification of ecotopes. To create this validity we can use biological entities as eco-indicators. Thus the Netherlands have been divided into 18 different land units (landscapes) using different combinations of breeding bird species (Kwak and Reyrink 1984).

It is a general opinion that landscape classification utilizes different levels of resolution according to purpose, and that it is pointless to adopt a standard and rigid classification. However, it is important to recognize the scale at which the classification is worked out. For more details on landscape classification see Zonneveld (1995).

The geographical context is also important. For example, the classification of an Apennine or Alpine landscape is quite different from that of a Padanian lowland landscape (Po valley, northern Italy). In the first case the topographic complexity creates so many climatic, soil and vegetation constraints that within a few kilometres can be found completely different habitat types (Fig. 1.17). In contrast, in the Padanian valley several kilometres are required to find the same level of contrast. Again, moving from the Padanian valley to the midwestern plains of the USA, even more kilometres are needed before we find the same environmental contrasts. The advantages of using a classification mainly consist in the possibility of comparing different studies in different sites.

The apparent anthropocentric classification adopted by different authors at the human scale can

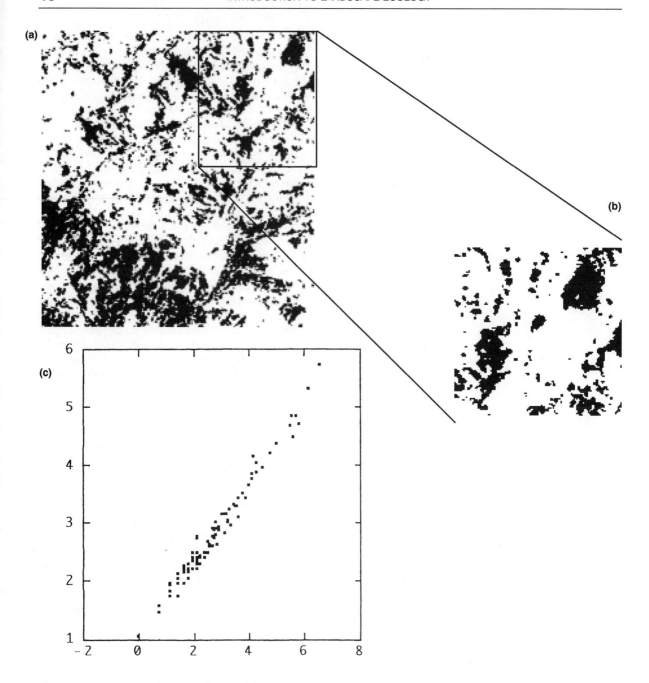

Figure 1.14 Patches are the elementary components of a landscape: shape and dimension are some of their main attributes. Example of patch analysis of a subset of a 2.5×2.5 km area (**b**) of a classified Landsat image, 8 × 8 km (**a**) (see Figure 1.11). Log of perimeter (*Y* axis) × log of area (*X* axis) is shown in (**c**).

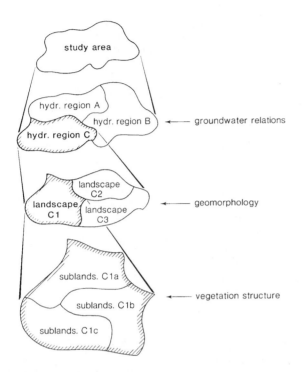

Figure 1.15 Example of landscape classification of the Netherlands based on a hierarchy of hydrological regions, geomorphology and vegetation structure (from Canters *et al.* 1991, with permission).

be easily converted when the landscape concept is used for vegetation and animal patterns. Instead of dogmatic conflicts in landscape ecology, often we find many convergences.

The hierarchy of anthropocentric landscape classification (ecotope, micro-, meso-, macro- and megachore) was created to describe the different processes that are of interest to people, comparable to the administrative organization of a region. An olive orchard may be compared to an ecotope, and the combination of olive orchard plus alfalfa field plus woodlot may be considered a microchore (land facet); this corresponds to a farm in the northern Apennines. A sequence of farms of this type, the so-called *coltura mista appenninica*, composes a mesochore (land system) that could represent a parish. The combination of mesochores creates a landscape (macrochore).

This classification is functional for human use and has a relevant ecological meaning, producing much information. These four levels of land classification link human socioeconomic structure to environmental resource allocation. This model works well in a rural perspective, but in an urban and industrial landscape the hierarchical factors change, losing the ecological (*sensu strictu*) feedback.

The information contained in this classification can be measured using species diversity (Whittaker 1977). Alpha diversity measures the microchore complexity, beta diversity the mesochore, gamma diversity the macrochore, delta diversity the complexity of a region, and finally epsilon diversity measures the complexity between regions (Naveh 1994).

An example of levels and type of species diversity is presented by Wiens (1989), who measured the diversity of birds according to different area aggregation. It is clear that these choices are arbitrary and depend on the goal of the study, but it is universally accepted that different levels of the ecological hierarchy have different information.

1.9 SUMMARY

- Landscape ecology is a young discipline that evolved in central and eastern Europe after World War II.
- The geobotanic, animal and human components of the landscape are all part of the discipline.
- Different theories, such as island theory, hierarchy theory, and models such as metapopulation and source–sink, have concurred to create the discipline.
- Different definitions are available.
- Space is the final frontier of the ecology theory.
- Behavioural ecology is strictly linked to landscape ecology.
- Patches are the emerging elements in the landscape.
- Physiotope, ecotope, mesochore, macrochore and megachore are some possibilities for a hierarchical functional classification of the landscape.

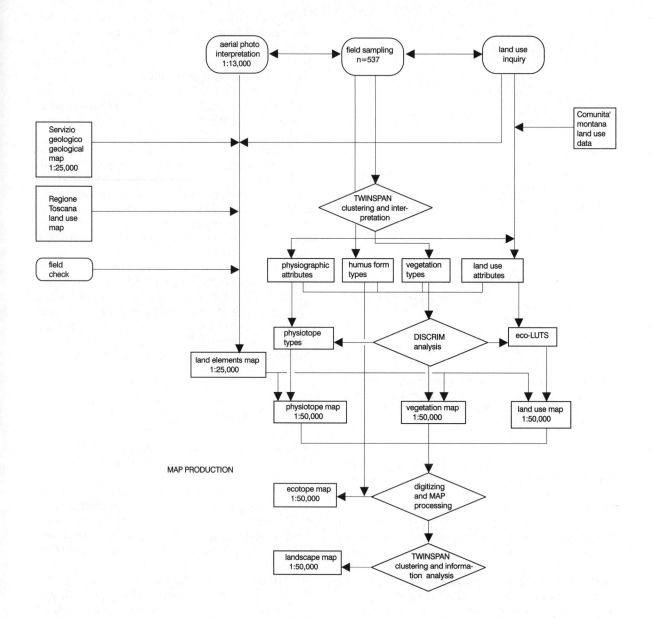

Figure 1.16 Example of a flow diagram of landscape classification and mapping (from Vos and Stortelder 1992, with permission).

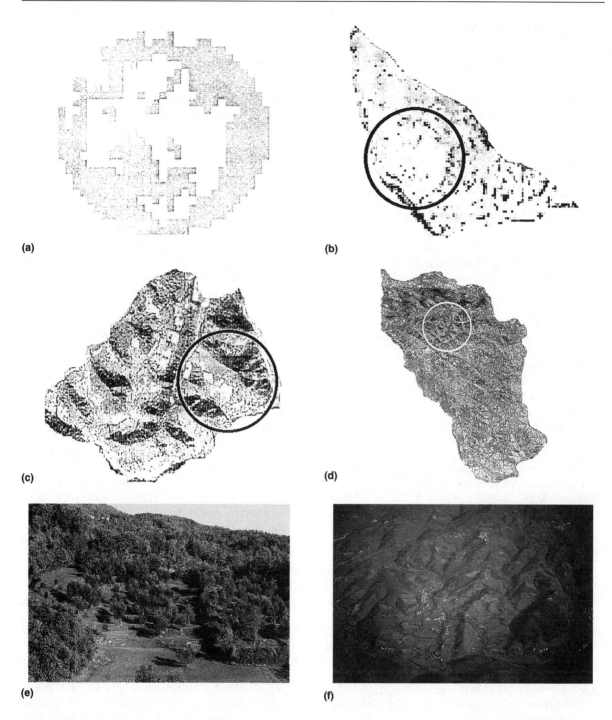

Figure 1.17 Example of landscape classification for a northern Apennines *coltura mista* landscape (the Olivola basin). (**a**) Ecotope represented by an olive orchard (100 m scale); (**b**) Microchore. A mosaic of olive orchards and *Quercus pubescens* located on the southern slopes); (**c**) Mesochore. A small stream catchment composed of a pattern of microchores; (**d**) Macrochore; (e) A detail of the olive orchards; (f) The Olivola basin represented by a Pliocenic lacustrine deposit heavily eroded and uplifted during the Quaternary.

SUGGESTED READING

The theory of island biogeography. MacArthur, R.H. and Wilson, E.O. Princeton University Press, Princeton, 1967.

Geographical ecology, patterns in the distribution of species. MacArthur, R.H. Princeton University Press, Princeton, 1972.

Hierarchy, perspectives for ecological complexity. Allen, T.H.F. and Starr, T.B. University of Chicago Press, Chicago, 1982.

Landscape ecology and land use. Vink, A.P.A. Longman, London and New York, 1983.

The fragmented forest, island biogeography theory and the preservation of biotic diversity. Harris, L.D. University of Chicago Press, Chicago, 1984.

Landscape ecology. Forman, R.T.T. and Godron, M. Wiley and Sons, New York, 1986.

Toward a unified ecology. Allen, T.H.F. and Hoekstra, T.W. Columbia University Press, New York, 1982.

Changing landscape: An ecological perspective. Zonneveld, I.S. and Forman, R.T.T. Springer Verlag, Berlin, 1990.

Landscape ecology. Theory and application, 2nd edn. Naveh, Z. and Lieberman, A.S. Springer-Verlag, New York, 1994.

Land mosaics. The ecology of landscapes and regions. Forman, R.T.T. Cambridge Academic Press, Cambridge, 1995.

Landscape ecology. Zonneveld, I.S. SPB Academic Publishing, Amsterdam, 1995.

Macroecology. Brown, J.H. University of Chicago Press, Chicago, 1995.

REFERENCES

Allen, T.H.F. and Starr, T.B. (1982) *Hierarchy, perspectives for ecological complexity.* University of Chicago Press, Chicago.

Blondel, J., Perret, P., Maistre, M., Dias, P.C. (1992) Do harlequin Mediterranean environments function as source–sink for blue tits (*Parus caeruleus* L.)? *Landscape Ecology* **6**: 212–219.

Buol, S., Hole, F.D., McCracken, R.J. (1989) *Soil genesis and classification,* 3rd edn. Iowa State University Press, Ames.

Canters, K.J., den Herder, C.P., de Veer, A.A., Veelenturf, P.W.M., de Waal, R.W. (1991) Landscape-ecological mapping of the Netherlands. *Landscape Ecology* **5**: 145–162.

Farina, A. (1993a) *L'ecologia dei sistemi ambientali.* CLEUP, Padova.

Farina, A. (1993b) Birds in a riparian landscape. In: Hagemeijr E.J.M. and Vestrael T.J. (eds.), *Bird Numbers 1992,* Statistics Netherlands, Voorburg/Heerlen, pp. 565–578.

Forestry Commission (1991) *Community woodland design.* HMSO Publication Centre, London.

Forman, R.T.T. (1995) *Land mosaics. The ecology of landscapes and regions.* Cambridge Academic Press, Cambridge, UK.

Forman, R.T.T. and Godron, M. (1986) *Landscape ecology.* Wiley and Sons, New York.

Gilpin, M. and Hanski, I. (eds.) (1991) *Metapopulation dynamics: empirical and theoretical investigations.* Academic Press, London.

Green, B.H., Simmons, E.A., Woltjer, I. (1996) *Landscape conservation. Some steps towards developing a new conservation dimension.* A draft report of the IUCN–CESP landscape Conservation Working Group. Department of Agriculture, Horticulture and Environment, Wye College, Ashford, Kent, UK.

Gross, J.E., Zank, C., Hobb, N.T., Spalinger, D.E. (1995) Movement cues for herbivores in spatially heterogeneous environments: responses to small scale pattern. *Landscape Ecology* **10**: 209–217.

Hansen, A.J. and Di Castri, F. (1992) *Landscape boundaries.* Springer-Verlag, New York.

Huggett, R. J. (1995) *Geoecology. An evolutionary approach.* Routledge, London.

Johnson, A.R., Wiens, J.A., Milne, B.T., Crist, T.O. (1992) Animal movements and population dynamics in heterogeneous landscapes. *Landscape Ecology* **7**: 63–75.

Kareiva, P. (1994) Space: the final frontier for ecological theory. Ecologyt **95**: 1

Kotliar, N.B. and Wiens, J.A. (1990) Multiple scales of patchiness and patch structure: a hierarchical framework for the study of heterogeneity. *Oikos* **59**: 253–260.

Kolasa, J. and Pickett, S.T.A. (1991) *Ecological heterogeneity.* Springer-Verlag, New York.

Kwak, R.G.M. and Reyrink, L.A.F. (1984) National breeding bird districts and their relation to landscape features. In: Brandt, J. and Agger, P. (eds.), *Methodology in landscape ecological research and planning.* Proc. First Int. Seminar IALE, Roskilde. Vol. III, pp. 27–39.

Lidicker, W.Z. Jr (1995) The landscape concept: something old, something new. In: Lidicker, W.Z. Jr (ed.) *Landscape approaches in mam-*

malian ecology and conservation. University of Minnesota Press, Minneapolis, pp. 3–19.

Lima, S.L. and Zollner, P.A. (1996) Towards a behavioural ecology of ecological landscapes. *TREE* **11**:131—135.

MacArthur, R.H. and Wilson, E.O. (1967) *The theory of island biogeography*. Princeton University Press, Princeton.

MacArthur, R.H. (1972) *Geographical ecology, patterns in the distribution of species*. Princeton University Press, Princeton.

Mandelbrot, B.B. (1975) *Les objects fractals: Forme, hasard et dimension*. Flammarion, Paris.

Merriam, G. (1984) Connectivity: a fundamental ecological characteristic of landscape pattern. In: Brandt, J. and Agger, P. (eds.), Proceedings of the First International Seminar on Methodology in Landscape Ecological Research and Planning. International Association for Landscape Ecology, Roskilde University Centre, Roskilde, Denmark, pp. 5–15.

Naveh, Z. (1987) Biocybernetic and thermodynamic perspectives of landscape functions and land use patterns. *Landscape Ecology* **1**: 75–83.

Naveh, Z. (1994) From biodiversity to ecodiversity: A landscape–ecology approach to conservation and restoration. *Restoration Ecology* **2**: 180–189.

Naveh, Z. and Lieberman, A.S. (1984) *Landscape ecology. Theory and application*. Springer-Verlag, New York.

O'Neill, R.V., DeAngelis, D.L., Waide, J.B., Allen, T.F.H. (1986) *A hierarchical concept of ecosystems*. Princeton University Press, Princeton.

Ostfeld, R.S. (1992) Small-mammal herbivores in a patchy environment: individual strategies and population responses. In: Hunter, M.D., Ohgushi, T., Price, P.W. (eds.), *Effects of resource distribution on animal–plant interaction*. Academic Press, San Diego, pp. 43–74.

Pearson, S.M. (1991) Food patches and the spacing of individual foragers. *Auk* **108**: 355–362.

Pickett, S.T.A. and White, P.S. (1985) *The ecology of natural disturbance and patch dynamics*. Academic Press, London

Pulliam, R. (1988) Sources–sinks, and population regulation. *American Naturalist* **132**: 652–661.

Risser, P.G., Karr, J.R., Forman, R.T.T. (1984) Landscape ecology. Directions and approaches. *Illinois Natural History Survey Special Publication number 2*, Champaign, Illinois.

Smith, A.T. and Peacock, M.M. (1990) Conspecific attraction and the determination of metapopulation colonization rate. *Conservation Biology* **4**: 320–323.

Southwick, E.E. and Buchmann, S.L. (1995) Effects of horizon landmarks on homing success in honey bees. *American Naturalist* **146**: 748–764.

Spies, T., Ripple, W.J., Bradshaw, G.A. (1994) Dynamics and pattern of a managed coniferous forest landscape in Oregon. *Ecological Applications* **4**: 555–568.

Swihart, R.K., Slade, N.A., Bergstrom, B.J. (1988) Relating body size to the rate of home range use in mammals. *Ecology* **69**: 393–399.

Turner, M.G. (1989) Landscape ecology: the effect of pattern on process. *Annual Review of Ecological Systems* **20**: 171–197.

Turner, M.G., Gardner, R.H., O'Neill, R.V. (1995) Ecological dynamics at broad scales. *Bioscience* S–29.

Vos, W. and Stortelder, A. (1992) *Vanishing Tuscan landscapes*. Pudoc Scientific Publishers, Wageningen.

Whittaker, R.H. (1977) Evolution of species diversity in land communities. *Evolutionary Biology* **10**: 1–67.

Wiens, J.A. (1989) *The ecology of bird communities*. Cambridge University Press, Cambridge, UK.

Wiens, J.A. and Milne, B.T. (1989) Scaling of landscapes in landscape ecology, or landscape ecology from a beetle's perspective. *Landscape Ecology* **3**: 87–96.

Woodmansee, R.G. (1990) Biogeochemical cycles and ecological hierarchies. In: Zonneveld, I.S. and Forman, R.T.T. (eds.) *Changing landscapes: An ecological perspective*. Springer-Verlag, Berlin, pp. 57–71.

Zonneveld, I.S. (1995) *Landscape ecology*. SPB Academic Publishing, Amsterdam

Theories and models incorporated into the landscape ecology framework 2

2.1 INTRODUCTION

The heterogeneity of the landscape, the complexity of the components of the system, the change in the behaviour of populations living in such an environment and the pressure of the habitat resource constraints have a strong effect on organisms in that landscape.

At least two theories (hierarchy and percolation) and two population models (metapopulation, source–sink) occupy an important place in the formulation of a landscape ecology framework. Despite having evolved in different contexts, these issues have in common the goal of interpreting the complexity and the heterogeneity of the environment (landscape). Their contribution to a homogeneous disciplinary body of landscape ecology is unquestionable, representing a paradigmatic bridge across the complexity of the landscape.

2.2 HIERARCHY THEORY AND THE STRUCTURE OF THE LANDSCAPE

Hierarchy is a very useful theory in landscape ecology for exploring many patterns and processes across different levels of spatiotemporal scales. Considering complexity as an intrinsic attribute of a landscape, the hierarchy paradigm explains how the different components localized at a certain scale are

in contact with the other components visible at a different scale of resolution.

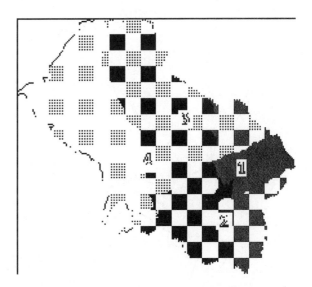

Figure 2.1 The river watersheds are examples of the hierarchical organization of the landscape. (1) Small-scale watershed (Rosaro stream, 2 km); (2) medium-scale watershed (Aulella river, 2 km); (3) semilarge-scale watershed (left watershed of the Magra river, 2 km); (4) large-scale watershed (upper basin of the Magra river, 2 km). The upper and lower limits of this hierarchy are not definitive but it is possible to move in both directions, including smaller and larger basins. (Magra river is located in northern Italy.)

The hierarchy theory considers a system as a component of a larger system, and composed in its turn of subsystems. Moving from one level to another of the system, the characters of phenomena change.

Landscape classification is one example of a hierarchical framework, moving from ecotope across micro-, meso-, macro- and megachores.

River watersheds are examples of a hierarchical system. A river basin is composed of sub-basins, each of which is composed of smaller-order basins (Fig. 2.1).

Complexity is a fundamental part of the hierarchy concept: the more components are included in a system, the more complex that system becomes. For this reason we can consider a landscape a very complex system (Fig. 2.2).

A system exists independently from his components and is generally self-organizing, so that it can be considered a cybernetic organization. Landscape has an organized complexity, and in order to understand a system it is necessary to focus on the level of organization. In fact, considering the complexity of a system it is important to select the best spatiotemporal scale at which the phenomena are related. The complexity of a system can be separated into vertical and horizontal structures.

With vertical structure we expect behaviour to occur at a slow rate. For example, leaves respond very quickly to light intensity, increasing photosynthesis, but the growth of a tree is represented by the integration of short time events. Not all vertical systems are hierarchical, but in a hierarchical system it is possible to isolate a layer according to different rates. Every layer of the system communicates with the others, filtering the messages that cross the border. A high frequency characterizes the lower layers of the system, but higher layers have processes occurring at a low frequency (Fig. 2.3).

The horizontal structure of a hierarchical system is composed of subsystems or holons. Holons may also be considered as an interface between the parts and the rest of the universe. Every holon is part of a higher-level holon, but can itself be considered as an assemblage of units. The boundary of the holon may be visible and tangible, such as the border of a forest, or intangible, such as the distribution of a population. One holon may be composed of other holons, which transmit an aggregated output to the higher-level holon. This represents a real filter for energy, material and information crossing the different layers. Hierarchy can also be defined as a system of communication in which holons with a slow behaviour are at the top and represent the context in which holons of lower level move faster.

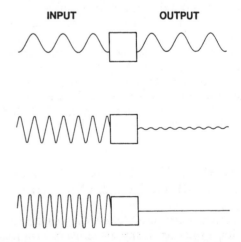

Figure 2.3 Changing of input signal when crossing a hypothetical constraint. Low-frequency signal crosses the box maintaining the characters unchanged; increasing input frequency produces a smoothing in the output. This behaviour is applicable to many natural processes across a hierarchical system (from O'Neill *et al.* 1986, with permission).

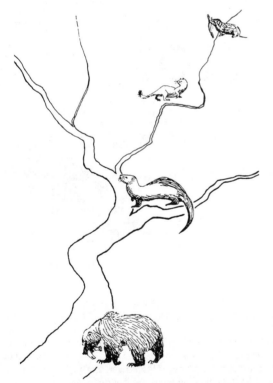

Figure 2.2 In a complex system organisms and processes may be structured in a spatial and functional hierarchy. In this case different-sized mammals are associated with stream order (from Harris 1984, with permission).

Scaling a hierarchical system consists in separating processes, and for this a forest may be considered a durable environment in which a species lives, or a dynamic entity, or simply a geomorphological process according to the scale of resolution available. The detection of a level is based on the rate of change of certain variables. Generally, where a discontinuity exists there may be a boundary of a hierarchical component.

Incorporation is a useful concept when we observe the behaviour of a system facing a perturbation or disturbance. This is defined as the process by which the perturbation is absorbed by a level of the system. Fire disturbance is a good example: a fire generally destroys part of a forest but not all of it, and the forest survives. In many cases fire is useful to ensure high diversity in the forest, and in this case fire is an incorporating disturbance.

Another good example is the grazing of ungulates. Grass cover suffers from the disturbance, but biomass consumption by the ungulates is replaced by a new stimulus of the root systems, thanks to nitrogen input released by manure deposition.

When the disturbance is very high and/or frequent the system collapses and complexity is reduced. An example of this is the coastal range of the Mediterranean basin, where human-induced fires occur so frequently that the system has not the capacity to incorporate the event. In this case transformation from forest to woodland then to scrubland represents progressive steps towards a simplification of the system. In this way some components of the overall system survive, but if the disturbance is severe the system is completely replaced by a different system.

The study of ecological complexity has long been overlooked by ecologists, but when we select the landscape scale composed of many ecological elements we cannot avoid considering the hierarchical arrangement of patterns and processes.

In studying an organism or a landscape we can consider the internal functioning of such an organism, and also its behaviour compared with external cues. In this case we have two levels, one higher and one lower. This is a pattern common to many natural organizations. Allen and Hoekstra (1992) distinguish five interrelated criteria ordering higher and lower levels:

1. Strong connections exist within the organism but only weak signals (energy, information) cross the surface.
2. Relative frequency. This is the number of times an organism repeats a behaviour, the frequency of which is determined by an internal clock. High-level systems have longer return times than small systems. There are many examples: for instance, large animals such as carnivores feed once or twice a day, but small animals such as shrews (only a few grams of biomass) need food every few minutes to compensate for their high metabolic rate (Fig. 2.4).
3. Context: the environment in which a lower level is contained.
4. Containment. In a nested system the higher level behaves more slowly than its parts, in which the whole is the context of the parts.
5. Constraint. Constraint may be considered as the limiting factor of a level. This point and the frequency are important criteria to order levels and allow systems to be predictable.

Recently Holling (1992) presented the hypothesis that a small set of plant, animal and abiotic processes structure ecosystems across scales in time and space. These processes have dominant temporal frequencies that control other processes and generally differ from each other at least by an order of magnitude. Therefore, we can expect these ecosystems to have few frequencies, endogenously driven and discontinuously distributed.

The discontinuity of frequencies is coupled with discontinuous distribution of spatial structures. So, animals living in such systems should have gaps in their size distribution according to the available landscape structure (Fig. 2.5).

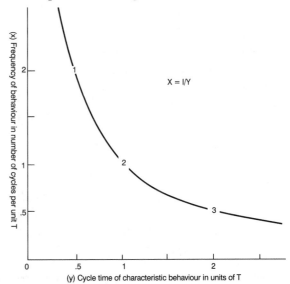

Figure 2.4 Graphical representation of inverse relationship between frequency of behaviour and cycle time (from Allen and Starr 1982, with permission).

Holling tested his hypothesis in different ecosystems (forest, grassland and marine pelagic) and on different animal groups having different body plans (birds and mammals) and feeding habits (carnivore, omnivore and herbivore). He found at least eight habitat 'quanta', each defined by a distinct texture at a specific range of scale, covering tens of centimetres to hundreds of kilometres in space, and from months to millennia in time.

The processes that influence structure move over a limited scale range.

Behaviour and the morphological attributes of animals can be used as a bioassay of landscape structure or as predictors of the impact of changes in vegetation pattern on animal community structures.

2.3 PERCOLATION THEORY

This theory, formulated to study the behaviour of fluid spreading randomly through a medium

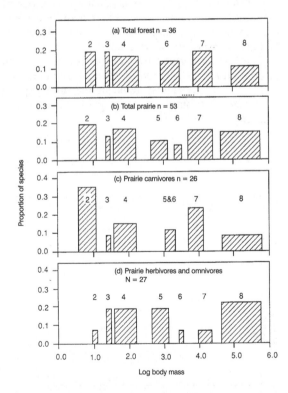

Figure 2.5 Distribution of mammals in boreal region forest and prairies according to the body-mass clump categories (from Holling 1992, with permission).

(Stauffer 1985), has found an interesting application in landscape ecology, in particular for the preparation of neutral models (Gardner *et al.* 1987).

Like the irregular thermal motions of molecules in liquid, in the diffusion process any diffusing particle can move and reach any position in the medium. The percolation process is quite different. A percolation threshold marks the differences between finite regions in which fluid remains when the percolation threshold pc is <0.5928 (also called critical probability) (Ziff 1986) or the fluid crosses the lattice connecting every molecule of fluid with the others, when P (probability) > pc (Fig. 2.6).

In large lattices the number and size of clusters are a function of P (probability that a cell is occupied by a target object, which can be a vegetation type or an animal distribution). The behaviour of the clusters changes rapidly around the critical threshold pc. If we measure the number of edge cells, the cells that are adjacent to an unoccupied map site, according to the P value it is possible to predict the amount of edge and of internal edges according to the fraction of the map occupied.

The importance of percolation theory in the study of landscape character is quite clear when we consider that near pc (= 0.5928) contagion effects, disturbance, forest fires and pest outbreaks have the starting point (Turner 1987). Percolation theory has been employed, for instance, in the study of landscape boundaries (Gardner *et al.* 1992). Considering a matrix composed of $m \times m$ cells, the extension of an ecotone across clusters depends on the probability P of occupying cells. Figure 2.6 illustrates three examples of P at 0.4, 0.6 and 0.8 of occupancy. The highest level of clusters is shown by the matrix with P occupancy of 0.4. No clusters are percolating. In case b a cluster is percolating and the number of clusters has been reduced to approximately half, compared with the first example. In Figure 2.6 only a cluster is present with a probability $P = 0.8$. According to this behaviour we can predict the amount of edge (total and inner) in the matrix. Figure 2.7 shows the amount of edges according to the fraction of map occupied.

Percolation theory also finds an application in the study of animal movements and the use of resources. When an animal moves in a habitat that has a value equal to or higher than $pc = 0.5928$, the organism can cross the entire landscape. Assuming that an organism can find at least one resource moving for n units of landscape, the probability of finding 0 resources is $(1 - P)^n$, where P is the random

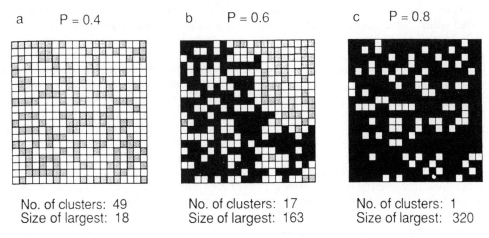

a   P = 0.4   b   P = 0.6   c   P = 0.8

No. of clusters: 49   No. of clusters: 17   No. of clusters: 1
Size of largest: 18   Size of largest: 163   Size of largest: 320

Figure 2.6 Example of three random maps with different percolation values: (**a**) no percolation, $P = 0.4$; (**b**) percolation, $P = 0.6$; (**c**) percolation, $P = 0.8$ (from Gardner *et al.* 1992, with permission).

distribution of the resource. The probability R of finding at least a resource is

$$R = 1 - (1 - P)^n \qquad (2.1)$$

We know from percolation theory that if $R = 0.5928$ then an organism can move from one part to another of the landscape. Substituting R in Equation 2.1 and rearranging it, we can find the relation between n and P:

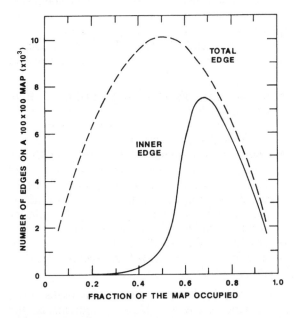

Figure 2.7 Number of edges in fraction of map occupied (p) (from Gardner *et al.* 1987, with permission).

$$n = -0.89845/\ln(1 - P) \qquad (2.2)$$

Equation 2.2 allows us to calculate the scale at which an organism interacts with the environment when the resources have a P distribution. If the resources are concentrated (P close to pc threshold) the number of landscape units that need to be explored by an organism is very low, but by decreasing the distribution of resources the number of units to be explored increases (Table 2.1).

If resources are two or more, the $n'' = -0.89845/(\ln(1 - P1) + \ln(1 - P2))$, where P1 and P2 are the distributions of the two resources.

Evidence of this mechanism is reported by O'Neill *et al.* (1988). When a dominant organism consumes 90% of the resources the subdominant organism has just 10% available. In this case, to

Table 2.1 Number of landscapes units (n) searched by an organism with a distribution of resources Pi (from O'Neill *et al.* 1988, with permission).

n	Pi
1	0.592800
4	0.201174
9	0.095007
16	0.054606
25	0.035300
100	0.009844
400	0.002244
900	0.000998
1600	0.000561
2500	0.000359

find the necessary amount organism has to move around to find other resources. As predicted by Equation 2.2, large-scale subdominant organisms are rare when sampled at a small scale.

2.4 THE METAPOPULATION

2.4.1 Introduction

The increase in forest fragmentation as a general trend in natural habitats has created small and isolated woodlots containing fewer species than surrounding woodlands. The reduction of species has been attributed to the high rate of extinction (Fig. 2.8). When populations living in a heterogeneous environment can be isolated from each other by hostile or less favourable habitats, contact between them is ensured only by emigration or immigration. The risk of local extinction and the probability of recolonization mainly depend on the ability to maintain an exchange of individuals.

The success of colonization depends on many factors, such as the capacity to disperse. These populations are generally considered as components of a metapopulation. Colonization ability is essential for the turnover process in a metapopulation.

The term metapopulation was introduced by Levins (1970) to describe a population of populations (Gilpin and Hanski 1991, Hanski and Gilpin 1991). Instead of focusing on a population, Levins considered a set of subpopulations actively in contact with each other. Metapopulations are systems in

which the rate of extinction and recolonization creates a flux of individuals that ensures genetic connectivity between the subpopulations. This is a very common condition in disturbed and fragmented habitas. The genetic or demographic connection is the necessary factor for creating a metapopulation, otherwise we are faced with separate populations. For this reason this definition and the demographic rules built around it have found great success during recent years.

The metapopulation concept is strongly related to island biogeography (MacArthur and Wilson 1967), considering both colonization and extinction as fundamental processes. In particular the inclusion of the metapopulation concept in landscape ecology contributes to a strong ecological synthesis.

Despite relatively common agreement as to the concept of metapopulation as a dynamic process of species distribution, many different opinions exist on the mechanisms that operate (Fig. 2.9). Most of these belong to population demography and have only a marginal interest for the landscape approach (see also Peltonen and Hanski 1991, Hill *et al.* 1996).

2.4.2 Dispersion

Although in the past reproduction and mortality were considered important and exclusive patterns of the population process, dispersion appears a very important factor governing the demographic and spatial structure of the metapopulation. Hansson (1991) has recently considered three main factors responsible for dispersion, presenting a review of dispersal mechanisms:

- **Economic threshold.** An individual moves from a patch when the level of food resources drops below a critical level. Dispersion is very common in temporary habitats.
- **Conflicts over resources.** Generally dispersion is necessary to escape competition for limited resources, such as food, breeding sites and water. This mechanism may be present in female versus male dominance, young against adults, inferior social categories against dominant ones.
- **Avoidance of inbreeding.** This may be a proximate or an ultimate factor, and seems to be density independent.

Hansson also considered the timing of dispersal, the genetic differences between individuals, the demographic differences and the spatial extent of dispersal as components of a complicated mecha-

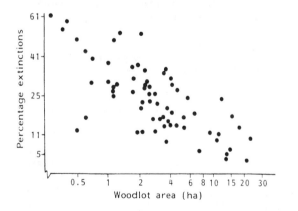

Figure 2.8 Effect of fragmentation on percentage of extinction of bird species in mature deciduous and mixed forest in farmland mosaic (from Opdam *et al.* 1994, with permission).

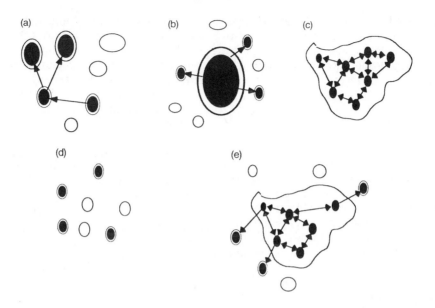

Figure 2.9 Different types of metapopulation models. Filled circles = occupied, unfilled = vacant; outlines represent the boundaries of a population; arrows indicate migration. (**a**) Levin's metapopulation; (**b**) core-satellite metapopulation; (**c**) patchy metapopulation; (**d**) non-equilibrium metapopulation; (**e**) a combination of C and B types (from Harrison 1991, with permission).

nism that moves individuals and populations in the constellation of metapopulations.

2.4.3 Examples of metapopulation structure

Melitea cinxia populations studied by Hanski *et al.* (1994) in the Baltic Aland island show strong evidence that long-term persistence largely depends on the 'genuine extinction–colonization dynamic'. An extensive migration was found that affects local density. The demographic and dynamic model of this species, in which subpopulations are not really very isolated, seems common for other species of butterflies.

In *Rana lessonae* (Gulve 1994) living in ponds the rate of extinction depends on deterministic and stochastic components. Deterministic extinction is mainly caused by ponds disappearing by drainage, or by natural succession. Extinction in permanent ponds depends on population stochastic processes, and the low rate of extinction found in permanent ponds (maximum 8.5%) strongly indicates that the correlation between ponds reduces the effect of the local extinction.

2.4.4 Metapopulation and conservation biology

The metapopulation model is extremely useful when applied to species conservation in a fragmented environment (Fig. 2.10). Recently Opdam *et al.*

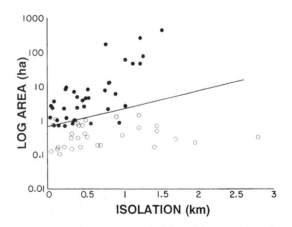

Figure 2.10 Effect of area size and island isolation on the presence (black dots) and absence (open circles) of the shrew *Sorex araneus* on islands in two lakes of Finland (from Peltonen and Hanski 1991, with permission).

(1994) discussed the landscape ecology approach as a basis for spatial planning. Fragmentation is dangerous when the patches are isolated and the metapopulation model cannot work. This point will be discussed further in Chapter 6.

2.5 THE SYSTEMS SOURCE–SINK

2.5.1 Definition

A source is a population in which births exceed deaths and emigration exceeds immigration. On the other hand, a sink population has a negative balance between offspring and death and juvenile production has not the capacity to compensate for adult mortality (Fig. 2.11). In the absence of immigration a population sink will face extinction.

Originally developed as a demographic model by Pulliam (1988), the source–sink paradigm finds full recognition, especially after the acceptance of the concept of heterogeneity and land mosaic complexity. It has recently been revised (Pulliam 1996) and conceptually enlarged (Dias 1996). We can consider a source patch as a place harbouring a source population and a sink patch the habitat occupied by a sink population.

Until the recent past, especially in population dynamics, most of the models were considered as homogeneous for every target habitat, without taking into account the spatial dimensions of the habi-

tats. In this way every individual experienced the same environmental conditions. But in reality various habitats inhabited by a species are not homogeneous in terms of resource availability, and for this reason are perceived as heterogeneous by individuals and subpopulations (Fig. 2.12).

This paradigm is extremely useful in landscape ecology to explain the different distribution of individuals across the mosaic. It is strictly linked to the metapopulation concept, having as a common basis the different conditions of the occupied patches and the interchange of individuals to maintain the system.

The quality of a patch is largely controlled by its size. In larger patches per capita production is greater and the source effect more evident. The reduction of source patches by fragmentation may have a serious effect on the survival of a population (Fig. 2.13).

2.5.2 Implications of the source–sink model

Allocation of a source or sink character to a patch or habitat is often difficult. For instance, a stochastic event particularly favourable to a species may occur in an unfavourable habitat for that species, and creates the wrong conviction that a particular habitat is of source type. For this reason particular attention should be paid before deciding the type of habitat. For this long-term studies are recommended.

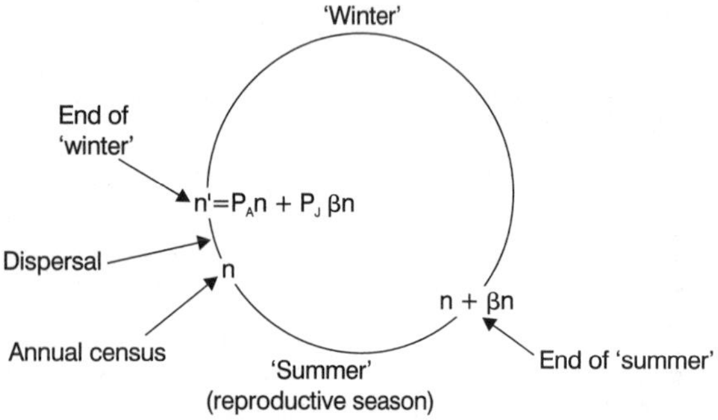

Figure 2.11 Demographic model in which at the end of the summer the population composed by n individuals has in total at the end of the reproductive season $n + \beta n$ individuals, where β are the juveniles alive at the end of the breeding season. At end of the winter the survival population n' is calculated by adult survival probability P_A and juvenile survival probability $P_J \beta n$ (from Pulliam 1988, with permission).

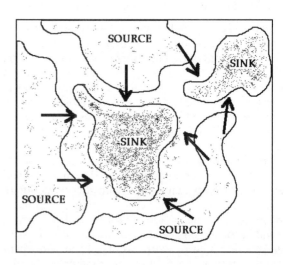

Figure 2.12 Representation of a source–sink model. Dots indicate the migratory flux from source to sink areas.

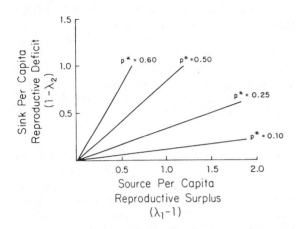

Figure 2.13 Graphical representation of a population at equilibrium in source patches (p) as function of source patch size and sink deficit (from Pulliam 1988, with permission).

2.5.3 Pseudo-sinks

Watkinson and Sutherlands (1995) coined the term pseudo-sink to illustrate the situation in which there are two habitats, of which one is more favourable than the other but with good carrying capacity. The poorer habitat is overpopulated because immigration exceeds the rate of birth/death. The difference between a true sink and a pseudo-sink is that if the sink is true the population becomes extinct if the immigration rate declines; in the case of a pseudo-sink the population will decrease if immigration is not active.

2.5.4 Traps

In some cases habitats appear extremely favourable to species although they have not the capacity to ensure conditions for a fully successful reproductive cycle. In other words, a trap is a sink that looks like a source (Pulliam 1996).

Such types of habitat are common in manmade landscapes in which, for example, food availability attracts a great number of individuals but the human disturbance linked to agricultural practices reduces reproductive success and consequently determines a demographic decline in a species. Such a habitat is consequently extremely harmful for some species. An example is the case reported by Pulliam (1996) of the grasshopper sparrows which in the southeastern United States are attracted by hayfields in early spring, but these cultivations are mowed in late spring or early summer before the sparrows have completed their breeding cycle.

2.5.5 Source–sink in time, or multiple source–sink model

The source–sink paradigm can find interesting applications in landscape ecology even when we consider migratory birds that change completely the quality of their habitat. Most trans-Saharan migratory birds have different breeding, wintering and migratory habitats. In this case the source–sink model appears difficult to apply, but could be extended to these intricate situations where we can distinguish in each separate seasonal habitat the characters of source or sink. One example is birds wintering in the Mediterranean region. In this case it would be possible to use this paradigm to define source wintering habitats as those in which the survival rate and the spring emigration are reasonably high. In contrast, a sink wintering habitat may be one that produces a progressive starvation in the animals, and consequently a very low overwintering success. Although difficult to measure, these conditions are very common in Mediterranean regions.

2.5.6 Stable maladaptation

A particular condition of maladaptation interpretable with the source–sink model has been presented by Blondel *et al.* (1992), studying the reproductive success of the bluetit (*Parus caeruleus*) breeding in deciduous (*Quercus pubescens*) and evergreen (*Quercus ilex*) oak forests of southern France (Fig. 2.14).

The populations breeding in deciduous oaks are well synchronized with food availability. In evergreen forest the laying date is the same but is not synchronized with the food availability, which occurs three weeks later. In this case the reproductive rate is lower. This phenomenon is explained by authors as a maladaptation of the birds that breed in evergreen forests but immigrate from the deciduous forest. In Corsica, where the evergreen forest is predominant and the bluetit populations are genetically separate from the mainland populations, synchronization is observed.

During the recent evolution of the forest cover in Europe an inversion of the source–sink system could be happening in Corsica. In such a case the change from deciduous oak cover to evergreen cover has

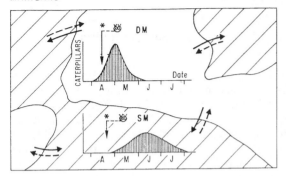

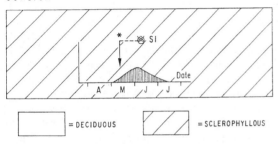

Figure 2.14 Example of source–sink system of bluetit (*Parus caeruleus*) in France and in Corsica. In mainland broadleaf forests the laying time is synchronized with food availability, but not in evergreen forests. In Corsica this species, genetically isolated from the mainland population, has its laying period synchronized within the evergreen forest. DM = deciduous mainland; SM = sclerophyllous mainland; SI = sclerophyllous island (from Blondel *et al.* 1992, with permission).

restricted the availability of deciduous oak forest and determined an evolutionary adaptation of the bluetit to this new habitat. In other words, when a sink habitat that generally should not be too much larger than a source habitat becomes predominant, it can cause evolutionary adaptation in a species.

2.5.7 Source–sink dynamics and conservation issues

Donovan *et al.* (1995a,b) have prepared a model to verify the effect of fragmentation on a source–sink system of neotropical migrant birds.

According to this model, which takes into account the metapopulation dynamic and the quality of the habitat patches, the fragmentation of

breeding habitats – a common pattern in the boreal regions – has a more significant effect in small habitat patches than in larger, core habitats. In sink habitats the decline may be higher in small patches and be independent of the source habitat (Fig. 2.15).

This assumption has found evidence and appears useful for managing endangered species. The fragmentation and the strong source–sink dynamic represent a cost for the population, determining a decline.

The source–sink dynamic requires interpretation over a longer period, although the sink dynamic may contribute to the overall population size and to assuring longevity. Sinks can contribute to gene exchange only when the immigration/emigration rate among subpopulations is significant.

2.5.8 Concluding remarks

It seems increasingly clear that the distribution of individuals often does not fit the availability of habitats. In such a case the sink habitat will face the extinction of a population, unless there is a continuous flux of immigrants. As pointed out by Pulliam (1996), suitable habitats are often unoccupied; density is not always an indicator of habitat quality; organisms often occur in unsuitable habitats; and in some cases for some populations most of the individuals occur in sink habitats. This idea is confirmed in part by island biogeography and in part by the metapopulation paradigm. The source–sink model, on the other hand, justifies the presence of species in unsuitable habitats. Finally, these two sets of theories explain most of the demographic dynamics of species.

2.6 SUMMARY

- Hierarchy and percolation theory, metapopulation and source–sink models strongly contribute to the formulation of a disciplinary body of landscape ecology.
- Landscapes are complex hierarchical systems and their dynamics depend on the scaled level of organization.
- Percolation theory is particularly important to explain the behaviour of systems around the critical threshold of cover ($pc = 0.5928$).
- The behaviour of organisms is strongly influenced by the pc value.

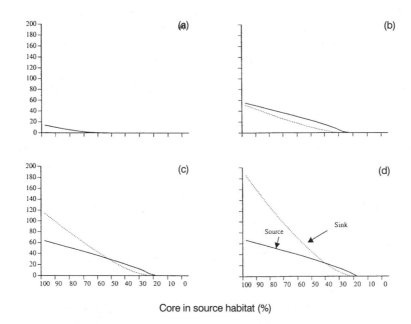

Figure 2.15 Population trends in source (solid line) and sink habitats (dotted line) compared with the percentage core in source habitats, at different rates of source fecundity. (**a**) Low, (**b**) observed, (**c**) medium, (**d**) high (from Donovan *et al.* 1995b, with permission).

- In fragmented and heterogeneous landscapes the subpopulations can be aggregated in metapopulations if the rate of exchange is ensured.
- A metapopulation reduces the risk of regional extinction.
- The source–sink model explains the different use of resources by organisms.
- Organism density is not always an indicator of source type.
- Multiple source–sink paradigms are of great utility in long-term nature conservation.

SUGGESTED READING

Gilpin, M. and Hanski, I. (eds.) (1991) *Metapopulation dynamics: empirical and theoretical investigations.* Academic Press, London.

MacArthur, R.H. and Wilson, E.O. (1967) *The theory of island biogeography.* Princeton University Press, Princeton.

O'Neill, R.V., DeAngelis, D.L., Waide, J.B., Allen, T.F.H. (1986) *A hierarchical concept of ecosystems.* Princeton University Press, Princeton, New Jersey.

Stauffer, D. (1985) *Introduction to percolation theory.* Taylor & Francis, London.

REFERENCES

Allen, T.F.H. and Starr, T.B. (1982) *Hierarchy. Perspectives for ecological complexity.* University of Chicago Press, Chicago.

Allen, T.F.H. and Hoekstra, T.W. (1992) *Toward a unified ecology.* Columbia University Press, New York.

Blondel, J., Perret, P., Maistre, M., Dias, P.C. (1992) Do harlequin Mediterranean environments function as source–sink for Blue Tits (*Parus caeruleus* L.)? *Landscape Ecology* **6**: 212–219.

Dias, P.C. (1996) Sources and sinks in population biology. *TREE* **11**: 326–330.

Donovan, T.M., Thompson III, F.R., Faaborg, J., Probst, J.R. (1995a) Reproductive success of migratory birds in habitat sources and sinks. *Conservation Biology* **9**: 1380–1395.

Donovan, T.M., Lamberson, R.H., Kimber, A., Thompson III, F.R., Faaborg, J. (1995b) Modeling the effects of habitat fragmentation on source and sink demography of neotropical migrant birds. *Conservation Biology* **9**: 1396–1407.

Gardner, R.H., Milne, B.T., Turner, M.G., O'Neill, R.V. (1987) Neutral models for the analysis of broad-scale landscape patterns. *Landscape Ecology* **1**: 19–28.

Gardner, R.H., Turner, M.G., Dale, V.H., O'Neill, R.V. (1992) A percolation model of ecological flows. In: Hansen, A.J. and di Castri, F. (eds.), *Landscape boundaries. Consequences for biotic diversity and ecological flows.* Springer-Verlag, New York, pp. 259–269.

Gilpin, M. and Hanski, I. (eds.) (1991) *Metapopulation dynamics: empirical and theoretical investigations.* Academic Press, London.

Gulve, P.S. (1994) Distribution and extinction patterns within a northern metapopulation of the pool frog, *Rana lessonae. Ecology* **75**: 1357–1367.

Hanski, I. and Gilpin, M. (1991) Metapopulation dynamics: brief history and conceptual domain. *Biological Journal of the Linnaean Society* **42**: 3–16.

Hanski, I., Kuussaari, M., Nieminen, M. (1994) Metapopulation structure and migration in the butterfly *Melitaea cinxia. Ecology* **75**: 747–762.

Hansson, L. (1991) Dispersal and connectivity in metapopulation. *Biological Journal of the Linnaean Society* **42**: 89–103.

Harris, L.D. (1984) *The fragmented forest. Island biogeography theory and the preservation of biotic diversity.* University of Chicago Press, Chicago.

Harrison, S. (1991) Local extinction in a metapopulation dynamics. *Biological Journal of the Linnaean Society* **42**: 73–88.

Hill, J.K., Thomas, C.D., Lewis, O.T. (1996) Effects of habitat size and isolation on dispersal by *Hesperia comma* butterflies: implications for metapopulation structure. *Journal of Animal Ecology* **65**: 725–735.

Holling, C.S. (1992) Cross-scale morphology, geometry, and dynamics of ecosystems. *Ecological Monographs* **62**: 447–502.

Levins, R. (1970) Extinction. In: Gertenshaubert, M. (ed.) *Some mathematical questions in biology. Lectures in mathematics in the life sciences.* American Mathematical Society, Providence, Rhode Island, pp. 77–107.

MacArthur, R.H. and Wilson, E.O. (1967) *The theory of island biogeography.* Princeton University Press, Princeton.

O'Neill, R.V., DeAngelis, D.L., Waide, J.B., Allen, T.F.H. (1986) *A hierarchical concept of ecosystems.* Princeton University Press, Princeton, New Jersey.

O'Neill, R.V., Milne, B.T., Turner, M.G., Gardner, R.II. (1988) Resource utilization and landscape pattern. *Landscape Ecology* **2**: 63–69.

Opdam, P., Foppen, R., Reijnen, R., Schotman, A. (1994) The landscape ecological approach in bird conservation: integrating the metapopulation concept into spatial planning. *Ibis* **137**: S139–S146.

Peltonen, A. and Hanski, I. (1991) Patterns of island occupancy explained by colonization and extinction rate in three species of shrew. *Ecology* **72**: 1698–1708.

Pulliam, H.R. (1988) Source–sink, and population regulation. *American Naturalist* **132**: 652–661.

Pulliam, H.R. (1996) Sources and sinks: empirical evidence and population consequences. In: Rhodes, O.E., Chesser, R.K., Smith, M.H. (eds.), *Population dynamics in ecological space and time.* University of Chicago Press, Chicago, pp. 45–69.

Stauffer, D. (1985) *Introduction of percolation theory.* Taylor and Francis, London.

Turner, M.G. (1987) Spatial simulation of landscape changes in Georgia: a comparison of 3 transition models. *Landscape Ecology* **1**: 29–36.

Watkinson, A.R. and Sutherland, W.J. (1995) Source, sinks and pseudo-sinks. *Journal of Animal Ecology* **64**: 126–130.

Ziff, R. (1986) Test of scaling exponents for percolation–cluster perimeters. *Physical Review Letters* **56**: 545–548.

Scaling patterns and processes across landscapes

3.1 INTRODUCTION

In cartography scale represents the level of reduction of the real dimensions of the earth, and may be absolute or relative. In ecology scale is a fundamental concept, as organisms interact with the environment using an inherent perception of their surroundings (Powell 1989, Steele 1989). Most ecological phenomena show a scale dependence of measurement and Horne and Schneider (1995) recently reviewed the role of spatial variance in ecology.

Although the scaling concept was for a long time used in ecological research, special emphasis has been devoted to it by plant ecologists (Cain 1943, Cain and Castro 1959, Greig-Smith 1964, Mueller-Dombois and Ellemberg 1974). More recently the concept of spatiotemporal scale has been used as a discriminatory element of complex processes such as extinction and recolonization, for the formulation of island biogeographical models (MacArthur and Wilson 1967).

Most ecologists consider scale to be an inherent property of the organism (Dayton and Tegner 1984, Wiens *et al.* 1986, Morris 1987, Maguire 1985, Bock 1987, Carlile *et al.* 1989), whereas others consider it a method to size a phenomenon without explicit influence on patterns and processes (Allen and Starr 1982, Maurer 1985). It is our opinion that both these assumptions can be accepted according to the context of the investigations (see later in this chapter), and we can adopt the principle that the scale most efficient to investigate a process is the one that allows us to collect the most information. In fact, moving across ecological processes, abiotic and biotic interactions have families of scales which exhibit emerging properties.

Often, observable patterns are determined by the collective behaviour of many small processes moving at different scales; in other conditions the patterns are induced by processes acting at large scales. For instance, in studying the flux of nutrients in kelp forests it is not enough to study processes at a local scale: these communities are concerned with processes that have scales of hundreds of kilometres. The type and the movements of sea currents occur on a broad scale, for example (Dayton and Tegner 1984).

3.2 MOVING ACROSS SCALES

The comprehension of processes that produce patterns is the essence of science, and many global and regional changes in biological diversity, pollutants, in greenhouse effects perceived at large scale, have their origins at the fine scale. Human influence is increasingly affecting patterns and processes at many different scales. Alteration and loss of habitat are more and more common, and this has tremendously depressing effects on biodiversity. To study the environmental problems of our planet it is necessary to create an interface between many factors that have different intrinsic scales.

To accept the movement across scales we must recognize the hierarchical organization of ecological systems. Moving top-down means to move toward an increase in detail. In contrast, moving bottom-up starts from the individual, across communities, ecosystems and the landscape. Moving across scales means accepting some levels of bias, due mainly to the heterogeneity that influences processes in a non-linear way. Often human biases are included in the research and in the interpretation of the results.

Recent advances in remote sensing and in geographical information systems have offered new opportunities for investigation at scales larger than in the past, and contemporary research into processes and patterns occurring in the microcosm has reconfirmed the importance of small scales. The expansion of spatial scale has led to a consequent enlargement of temporal scale which, in tracking back processes, has allowed us to understand the environmental conditions of the past (Delcourt and Delcourt 1988). The availability of data across scales has opened new possibilities for integrating patterns and processes, as recently stressed by Lubchenco *et al.* (1991).

There are many scales of interest, and at each scale certain processes are visible for their preeminent characters.

Generally phenomena are studied using a deliberate scale, but often the scale is chosen according to perceptual capabilities or by technological or logistical constraints (Magnuson 1990, Swansson and Sparks 1990, Wiens 1992, Walker *et al.* 1993).

We define the observation scale as one that we can use to measure a process, and the process scale or inherent scale is that exhibited by natural phenomena and independent of our control.

All organisms are conditioned by the intrinsic scale of resolution and generally have the ability to change this scale according to different situations. For instance, adult animals have a greater capacity to explore allocated resources than do juveniles, because they utilize a broader range of behaviours.

Dispersal and dormancy are two strategies to change the scale of perception.

Knowing what information is preserved and what is lost as one moves from one scale to another is of great interest. Moving across a scale, we can carry forward information from the fine scale to a broad scale, losing details or heterogeneity but gaining in predictability.

The scaling capacities of organisms allow them to maintain spatial and temporal patterns, with consequences for the dynamics of populations and ecosystems. Every species experiences the environment on a specific range of scales, responding individualistically to environmental variability. Local unpredictability and variability allow species to reduce the competition more than in a constant environment.

Although the individual locally has deterministic replies to environmental constraints, at the population level stochastic variables may come into play and the observed patterns receive an irrelevant contribution from the fine-scale behaviour. The choice of sample size and the resolution of quantitative analysis are two sides of the same coin.

3.3 DEFINITIONS

The word scale has various meanings in a number of disciplines. In ecology, and particularly in landscape ecology, scale refers to the spatial or temporal dimension in which an organism or a pattern or process is recognizable. Table 3.1 and Figure 3.1 give more details on terminology and definitions (Turner *et al.* 1989a). Changing scale in the analysis of a land mosaic means changing the resolution of cells or increasing the area of survey. A scale may be defined as the period of time or space over which signals are integrated or smoothed to give a message (Allen and Starr 1982).

The use of trivial attributes to describe organisms and their behaviour is often related to human perception. Size, reproductive time, longevity and rates of movements are intrinsic scaled factors. Extrinsic factors are the physical environment and human perception.

3.4 SCALING THE LANDSCAPE

In this chapter we emphasize the importance of scale for studying patterns and processes and the need to

Table 3.1   Some scale-related terminologies and concepts (from Turner *et al.* 1989a, with permission)

Term	Definition
Scale	The spatial or temporal dimension of an object or process, characterized by both grain and extent
Level of organization	The place within a biotic hierarchy (e.g. organism, deme, population)
Cartographic scale	The degree of spatial reduction indicating the length used to represent a larger unit of measure; ratio of distance on a map to distance on the earth surface represented by the map, usually expressed in terms such as 1:10 000
Resolution	Precision of measurement: grain size, if spatial
Grain	The finest level of spatial resolution possible with a given data set, e.g. pixel size for raster data
Extent	The size of the study area or the duration of time under consideration
Extrapolate	To infer from known values; to estimate a value from conditions of the argument not used in the process of estimation; to transfer information (a) from one scale to another (either grain size or extent) or (b) from one system (or data set) to another system at the same scale
Critical threshold	The point at which there is an abrupt change in a quality, property, phenomenon
Absolute scale	The actual distance, direction, shape and geometry
Relative scale	A transformation of absolute scale to a scale that describes the relative distance, direction, or geometry based on some functional relationship (e.g. the relative distance between two locations based on the effort required by an organism to move between them)

track phenomena across scales, taking into account that in landscape ecology the hierarchy theory is very popular and largely accepted (Meentmeyer and Box 1987, O'Neill *et al.* 1989, 1991).

Scaling physical processes such as runoff, the spatial distribution of plants and the behaviour of animals is an activity that considers the ecosphere as a hierarchical system in which patterns and processes moving from one layer to another modify their properties.

In landscape ecology more than in any other field of ecological research scale is a central point around which most of the investigations have to focus, and has the capacity to unify population biology and ecosystem science (Levin 1992). Landscape ecological research is posed from a scale of a few metres up to a thousand kilometres, across which most ecological processes are completed. The same applies to the temporal scale, which is often considered from the seasonal aspect, such as in the study of metapopulation dynamics, to the millennia resolution of biome modification.

A system functions across a variety of scales, and when observed at one resolution we perceive certain characteristics filtering most of the noise, owing to the close layering (sub and upper levels) of the entire organization.

Different methods from spatial statistics can be used to study the variation of processes according to scale, for instance fractals, semivariograms, correlograms, spectral analysis etc.

Patchiness and variability are exhibited by every population across a broad range of scales, and such a system is strategic for maintaining ephemeral or competitively inferior species which depend upon the local modification of resource availability and interspecies competition.

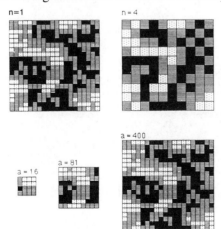

Figure 3.1 Two components of spatial scale: grain size and extent. (**a**) The grain size means the minimum dimension of elementary component (cell, pixel). In the figure the grain has been enlarged from 1 to 4. (**b**) The extent is the area considered. The increasing rate of the example is from 16 to 400 cells or pixels (from Turner *et al.* 1989b, with permission).

3.5 CHANGE OF PERCEPTION SCALE

If scale is considered an inherent character of species we could believe that every species perceives and reacts to its environment at the same scaling metric. But the capacity to change the scale of observation is typical not only of the human mind: many species have the ability to change their environmental perception, for instance by dormancy and by dispersion. Often the scale issue is more complicated than expected. In fact, most species have the capacity to change scale according to the seasons or their internal rhythms. The coarse or fine perception of the environment can change with the season, for instance. Some species of birds (e.g. the robin, *Erithacus rubecula*), recognize a site as coarse grained during the breeding season and fine grained outside the reproductive period. During the breeding season the robin selects woodlands completely avoiding shrubby and open areas. In this case behaviour is coarse grained, but outside this period the robin selects many different habitats (woody + shrubby + open), having a fine-grained perception of the environment (Farina 1996b).

Changing scale in the study of animal distribution allows us to understand the proximate factor respon-

sible for abundance. As illustrated in Figure 3.2, the distribution of robins across three spatial scales allows us to understand how a species shares the different habitats, the subregions and the regions. In the case represented in this figure the local scale (Fig. 3.2a) indicates how pairs are distributed in the habitat. At the catchment scale (Fig. 3.2b) the distribution of breeding birds indicates the variability in habitat suitability across a relatively restricted range of habitat availability. At the regional scale (Fig. 3.2c) the effect of slope orientation and the different climatic regime are important factors affecting the distribution of this species. Crossing different spatial scales allows investigation of nest site selection, habitat selection and regional selection.

3.6 THE MULTISCALE OPTION

Many animals interact with the environment using different scales. For instance, in *Apodemus sylvaticus* (Muridi, Mammipheres) and *A. flavicollis*

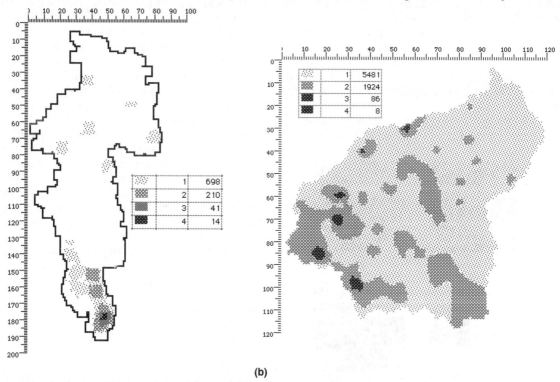

(a)   (b)

Figure 3.2 Distribution of breeding robin (*Erithacus rubecula*) across different geographical scales: (**a**) local scale (Logarghena prairies) grain size 20×20 km, extension 2×4 km (Farina 1996a); (**b**) catchment scale (Aulella river), grain size 200×200 m, extension 24×24 m (Farina 1996b); (**c**) regional scale (transect across the northern Apennines), grain size 200×200, extension 38×40 km (Nardelli 1996).

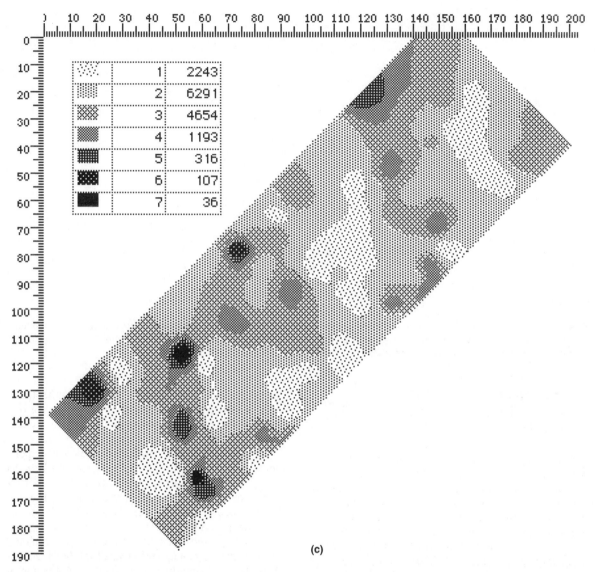

		1	2243
		2	6291
		3	4654
		4	1193
		5	316
		6	107
		7	36

(c)

(Muridi, Mammipheres) the environmental choice of habitat is linked by two mechanisms (Angelstam *et al*. 1987): at the microscale food availability seems to be the most important factor, but at the macroscale the discriminant component is the percentage of rural fields.

Carlile *et al*. (1989) argued that vegetation patterns depend on the overlap of factors such as soil composition, the nutrients, the humidity, the topography etc., which have a strong effect on vegetation. For this reason neighbouring plots should have a lower variance than more distant plots, varying the frequency of sampling along transects and the distance between transects. For instance, Carlile *et al*.

(1989) found that sample variance increased as a function of transect segment length, and intersegment length and correlation decreased as a function of intersegment length and transect segment length. For this reason *Agropyron spicatum* is correlated across the overlap on a scale of 400–700 m, and a segment length of 64–128 m is an appropriate scale to measure the cover of this plant (Fig. 3.3).

Krill and phytoplankton represent an example of multiscale processes. At the large scale are the sea currents that determine the location of these organisms, but it is at the fine scale that the diving behaviour of the krill affects the distribution and abundance of the phytoplankton.

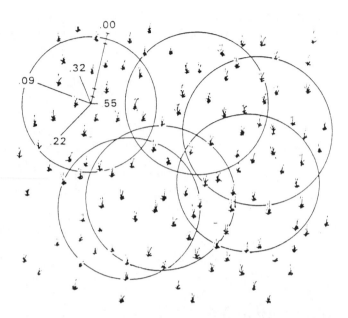

Figure 3.3 Spatial model of *Agropyron spicatum*. Circles represent inherent units of ecological scale. Any point in the landscape is the centre of a natural scale unit. The distance from each centre is measured and when the correlation declines to zero indicates the bound of the inherent unit of scale (from Carlile *et al.* 1989, with permission).

3.7 IMPORTANCE OF THE PARAMETERS AT THE DIFFERENT SCALES

Certain variables cannot change the scale, but it is their importance that changes. In this way, predicting the rate of decomposition in the soil means knowing the environmental variability, the microclimate and the characteristics of the litter. At a regional scale temperature and humidity are good predictors of decomposition rate. Evapotranspiration is controlled at the leaf or tree scale by the deficit of the steam and by the stomata processes, but at a regional scale solar radiation is the environmental variable.

The rate of mortality of oaks at a local scale has been found to decrease as rainfall increases. At the regional scale this mortality is lower in more arid regions.

3.8 GRAIN SIZE AND SCALING

Individuals perceive the environment as coarse or fine grained according to inherent characteristics. Generally, plants and other sessile organisms that spend most of their life in the same place have a coarse-grained perception of the environment, but when we distinguish competition processes for dispersal it is possible to recognize different scaling factors also in sessile organisms.

Stratton (1995) has conducted an elegant experiment on variation in fitness of *Erigeron annuus* (Fig. 3.4). This species shows genotype–environment interactions on small spatial scales, although the evidence of such a mechanism at 10–20 cm has been confirmed at 3 m intervals in only one experiment (1995), and not in the experiment conducted by the same author in 1994 (Stratton 1994). In this plant competition appears at 20 cm distance and at 3 m appears the process of dispersal (Fig. 3.5).

3.9 ASSESSING LANDSCAPE SCALE OF ANALYSIS

Scale is defined as a spatiotemporal dimension that produces the best information in the most efficient and unbiased way (Carlile *et al.* 1989). Wiens (1986) and Wiens *et al.* (1986) have argued that for studying interspecific competition the most appropriate scale should be the local or regional, but not the biogeographical scale.

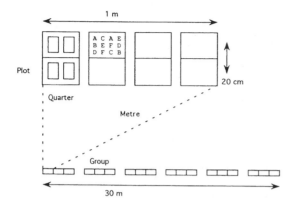

Figure 3.4 Experimental design of spatial arrangement of eight genotypes of *Erigeron annuus*. A transect of 30 m was divided into six groups, 3 m long and separated by a 2 m gap. Each group was divided into three quarters, each composed of two plots. Every plot was subdivided into two subplots of 5×10 cm. Every genotype was planted in each subplot. Two transects (10 m apart) were selected (from Stratton 1995, with permission).

In a theoretical sense the best should be achieved using a hierarchy of scales having the capacity to correlate abiotic, biotic and human processes (Krummel *et al.* 1987). According to Morris (1987), the selection of a scale should take into account the biological attributes of a species, such as home range, population density, dispersal movements etc. In this way the human scaling biases would be removed.

Efficient sampling of landscape characters often requires splitting the investigation into a scaled context – macro and micro habitats – that represent the grain of resolution (Fig. 3.6).

When landscape studies require different aggregations of data it often becomes difficult to select the right dimension at which patterns are visible and related to processes. The effect of changes of scale on the results is often underestimated or not easy to detect. For instance, remote sensing data actually available are mostly based on four main sources, with different resolutions:

SPOT	20×20 m
Thematic mapper	30×30 m
MSS	80×80 m
AVHRR	1×1 km

It is imperative to understand what happens when we move from one resolution to another. Qi

and Wu (1996) have tested three spatial autocorrelation indices (Moran coefficient, Geary ratio and Cliff–Ord statistic) to the topography and biomass of peninsular Malaysia in 1992. They found relevant effects on results from changing scale for all the three indices. This means that for some attributes of the landscape there are threshold levels, for instance for elevation, but this is not true for biomass, which seems less sensitive to the scale of resolution.

The influence of different spatial scales has been studied by Fuhlendorf and Smeins (1996) on the dynamics of common curly mesquite (*Hilaria belangeri* (Steud.)) (Table 3.2).

Large scale is characterized by low variations between sampling units, high variation within units, high predictability or equilibrium. At small scale variation between units is high, predictability low, and there are no indications as to the potential for evolution toward a stable or chaotic behaviour.

When we study landscape patterns the choice of resolution and extent is of fundamental importance to reduce bias. So, O'Neill *et al.* (1996) suggest that the grain should be 2–5 times smaller than the spatial features of interest, and that the sample area must be 2–5 times larger than landscape patches to avoid bias when indices of landscape structure, such as dominance, shape, contagion etc., are applied.

3.10 EXAMPLES OF SCALES IN LANDSCAPE AND IN ECOLOGICALLY RELATED DISCIPLINES

3.10.1 Scaling the Quaternary landscape

Most of the recent history and development of landscape is directly connected with human evolution. In particular, during the last 10 000 years humans and the landscape have evolved closely by feedback between natural and human-driven processes. For this reason the history of the recent past assumes a special value in landscape ecology. Changes in the structure and functioning of landscape, from hunting and gathering (protohistory) to the discovery of agriculture and the birth of permanent settlements, have affected most of our living realm (Fig. 3.7). Delcourt and Delcourt (1988) define four levels of scale related to landscape ecology issues to study the Quaternary landscape:

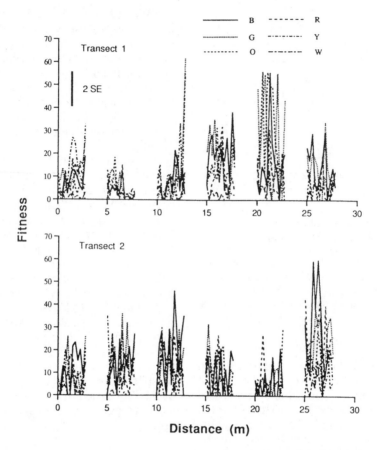

Figure 3.5 Spatial variation in the mean fitness of the six genotypes of *Erigeron annuus* along two transects. B, G, O, R, Y, W are the six genotypes. Fitness was estimated as the product of survival and the number of capitula. See Figure 3.4 for the experimental design (from Stratton 1995, with permission).

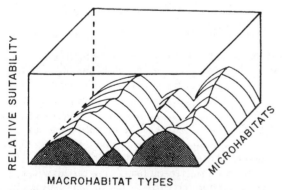

Figure 3.6 Macro and micro scales to measure habitat quality. Using a coarse approach we can observe different macrohabitat variability across a spatial distribution. Using a fine resolution it is the habitat quality and micro-site that are recorded (from Morris 1985, with permission).

- **Microscale dominion**. This scale considers a period from 1 to 500 years and a space from 1 m^2 to 10^6 m^2. Scientists working at this scale are geomorphologists, plant succession and animal ecologists and planners.

  - Disturbances such as fires, windthrow and clearcutting are involved at this scale;
  - Geomorphic processes, such as soil creep, movement of sand dunes, debris avalanches, slumps, fluvial transport and exposition, and cryoturbation;
  - Biological processes, such as the cycles of animal populations, gap-phase replacement in the forests, succession after abandonment;
  - In forested landscape fragmentation, increase of ecotones and change in corridor availability.

Table 3.2 Temporal dynamics of a grassland community dominated by common curly mesquite (*Hilaria belangeri*) according to different spatial scales (from Fuhlendorf and Smeins 1996, with permission)

	Scale	
	Large (exclosure)	*Small (quadrant)*
Variability between units	Low	High
Variability within unit	High	Unknown
Potential predictability	High	Low
Probability of equilibrium	Possible	Minimal
Probability of chaos	Minimal	Possible
Event of driving process	Large	Small (variable)

- **Mesoscale dominion.** The meso-scale dominion extends from 500 to 10 000 years and in space from 10^6 to 10^{10} m². In this period are comprised the events that range from the last interglacial interval, and in space from the watershed on second-order rivers. In this domain the cultural evolution of humans occurred.
- **Macroscale dominion.** The macro-scale dominion extends from 10 000 to 1 000 000

years, with spatial extension from 10^{10} to 10^{12} m². In this dominion the glacial–interglacial cycles occurred and speciation and extinction operated.
- **Megascale dominion.** This dominion extends from 10^6 to 4.6 billion years, with an extension $>10^{12}$ m², covering the American continent and interacting geological events such as plate tectonic movements.

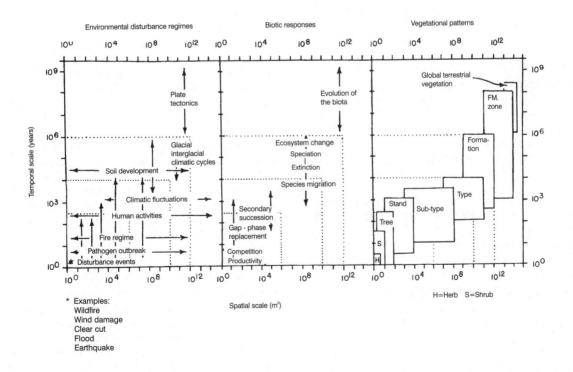

Figure 3.7 Effects of different disturbance regimes, biological responses and vegetational patterns in three spatiotemporal domains (from Delcourt and Delcourt 1988, with permission).

3.10.2 Scaling patterns: the catchment scale

This scale is very popular in these times. The catchment scale that pertains to terrestrial and aquatic ecosystems, recently reviewed by Hornung and Reynolds (1995), seems a very promising dimension in which the fluxes of water and elements link the different components of the systems. Disturbance regimes, such as agricultural intensification, afforestation and fire, can be monitored using the chemical composition of streams and underground and surface waters. In particular, small catchments can be adapted to study pollution, land management activity and environmental changes. Increasing spatial and temporal resolution is a further possibility for investigation in greater depth.

3.10.3 Scaling abiotic processes: hydrological processes and scales

Hydrological processes are a principal component of landscape mechanisms. Their dynamic has a dramatic influence on most of the abiotic and biotic processes. They range in eight orders of magnitude in space and time, occurring at a wide range of scales, from unsaturated flow 1 m soil profile to floods in river systems of a million square kilometres; from flash floods lasting some minutes to flow in aquifers over hundreds of years (see Bloschl and Sivapalan 1995 for a review) (Fig. 3.8).

Usually three levels of scale are used in the study of hydrological processes: (a) the lifetime (duration) (for an intermittent process such as a flood); (b) the period or cycle (for a periodic process such as snow-melt); and (c) the correlation length (integral scale), which represents the average distance of correlation between two variables.

In the hydrological studies that are basic components of the functioning of a landscape the space scale ranges from 1 m (local scale) to hillslope (reach) (100 m) to catchment scale (10 km), to regional scale (1000 km). In time we have three levels: the event scale (1 day), the seasonal scale (1 year) and long-term scale (100 years) (Fig. 3.9).

3.10.4 Scaling evidence in animals

Every species perceives the surrounding environment (landscape) in a different way. The movement of a grass stem appears dramatically great for aphids but is not perceived by a deer.

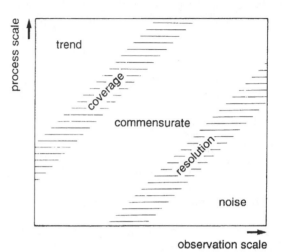

Figure 3.8 Plotting process scale with observation scale creates three regions in the design according to the sampling resolution. If the coverage is smaller than the process measured the information acquired can be described as a trend. In contrast, if the process is smaller, then the resolution appears as noise. The intermediate belt appears as a commensurate space (from Bloschl and Sivapalan 1995, with permission).

A landscape homogeneous for a species, such as a mountain prairie for the water pipit (*Anthus spinoletta* (Motacillidae, Aves)), is perceived as heterogeneous for *Erebia* sp. (butterfly, Satiridae, Lepidoptera), which distinguishes in this prairie patches with different food resource availabilities (Fig. 3.10).

Unfortunately often we select scales more comfortable to our metric than to that of the species in question (Dale *et al.* 1989, Wiens and Milne 1989, Wiens 1992), although it is possible to judge the best range of scales at which an organism spends its life (territorial behaviour, dispersal movements, food research) (O'Neill *et al.* 1988).

Animal behaviour interacts with environmental patterns and processes across several spatiotemporal scales (Gardner *et al.* 1989). That organisms have the capacity to move across a range of scales is well documented. This capacity is important, considering the complexity of the lifecycles of some organisms, such as mammals.

Dorcas gazelles (*Gazella dorcas*) foraging in the Negev desert (Ward and Saltz 1994) select a plot with a high density of madonna lilies (*Pancratium sickenbergeri*) and in this plot, at a small scale,

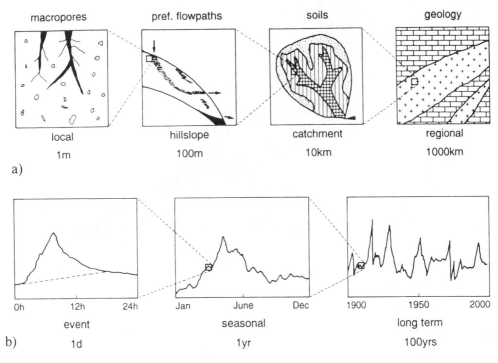

Figure 3.9 Component of catchment and hydrological processes across spatial and temporal scales (from Bloschl and Sivapalan 1995, with permission).

stays longer than in a random-walk model, but at a broad scale they move directly from one plot to another, suggesting that this species repeatedly samples the environment.

Many organisms are selected as bioindicators and for this they must be independent of the chosen scale. To test this hypothesis Weaver (1995) studied an arthropod community across a range of four spatial scales, each nested in the others. He found that the proportion of species in a sample depended on the scale of observation, but Acari and Collebola showed a decrease in richness upon increasing the scale of sampling, presenting a uniform distribution in the soil. The diversity of Aranaceae, Thysanoptera, Formicidae and Coleoptera larvae increased with the addition of new samples. In particular, Coleoptera added new species, from 4% in a sample to 22% of total species moving from samples ($n = 93$), plots ($n = 24$) to stand ($n = 4$). This means that these species are patchily distributed and also perceive the environment as patchy. Diptera larvae remained constant across the sample scale.

These data are relevant for monitoring biodiversity. If for many groups richness is a matter of spatial scale, to improve monitoring efficiency a spatial scale of sampling seems to improve the inventory results.

3.11 SUMMARY

- Scaling in landscape ecology allows us to link landscape dynamics, biodiversity and ecosystem processes.
- Organisms and processes interact across inherent scales.
- Investigation across scales allows us to track processes that are functioning along hierarchical systems.
- Scaling in landscape ecology is of great importance for the complexity of patterns and processes belonging to the landscape.
- Grain size and extension are components of scale attributes.

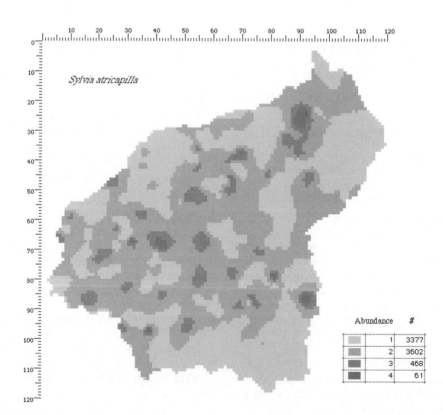

Sylvia atricapilla

Abundance	#
1	3377
2	3602
3	468
4	61

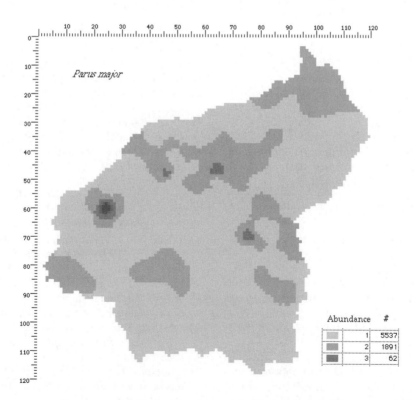

Figure 3.10 The perception of the landscape is species specific, so a small passerine bird perceives as patchy an environment with different patchy characters. The great tit (*Parus major*) perceives the Aulella landscape as coarser than the blackcap (*Sylvia atricapilla*). This patchiness perception is achieved by using the density isolines (from Farina 1997, with permission).

- The temporal scale is an important component in studying the evolution and dynamics of a system, especially during the last 10 000 years.
- Often organisms utilize a multiscale perception of the environment, selecting proximate to ultimate factors.
- Catchment scale and hydrological fluxes have scale components which are strategic to an understanding of the functioning of the landscape.
- Animals exhibit a capacity to change inherent scale according to different physiological status, avoiding competition or exploiting new resources.
- Animals can change their perception of the environment according to seasonal requirements.

SUGGESTED READING

Allen, T.F.H. and Starr, T.B. (1982) *Hierarchy. Perspectives for ecological complexity*. University of Chicago Press, Chicago.

Levin, S.A. (1992) The problem of pattern and scale in ecology. *Ecology* **7**: 1943–1967.

REFERENCES

Allen, T.F.H. and Starr, T.B. (1982) *Hierarchy. Perspectives for ecological complexity*. University of Chicago Press, Chicago.

Angelstam, P., Hansson, L., Pehrsson, S. (1987) Distribution borders of field mice Apodemus: the importance of seed abundance and landscape composition. *Oikos* **50**: 123–130

Bloschl, G. and Sivapalan, M. (1995) Scale issues in hydrological modelling: a review. *Hydrological Processes* **9**: 251–290.

Bock, C.E. (1987) Distribution–abundance relationship of some Arizona landbirds: a matter of scale? *Ecology* **68**: 124–129.

Cain, S.A. (1943) Sample plot technique applied to Alpine vegetation in Wyoming. *American Journal of Botany* **30**: 240–247.

Cain, S.A. and Castro, G.M. (1959) *Manual of vegetation analysis*. Harper, New York.

Carlile, D.W., Skalki, J.R., Batker, J.E., Thomas, J.M., Cullinan, V.I. (1989) Determination of ecological scale. *Landscape Ecology* **2**: 203–213.

Dale, V., Gardner, R.H., Turner, M.G. (1989) Predicting across scales: theory development and testing. *Landscape Ecology* **3**:147–252

Dayton, P.K. and Tegner, M.J. (1984) The importance of scale in community ecology: a kelp forest example with terrestrial analogs. In: Price, P.W., Slobodchinof, C.N., Gaud, W.S. (eds.). *A new ecology*. John Wiley and Sons, New York, pp. 457–481.

Delcourt, H.R. and Delcourt, P.A. (1988) Quaternary landscape ecology: relevant scales in space and time. *Landscape Ecology* **2**: 23–44.

Farina, A. (1996a) Diversity and structure of bird assemblages in upland pastures facing land abandonment. Proceedings of Fifth European Forum on Nature Conservation and Pastoralism, Cogne, 17–21 September 1996 (in press).

Farina, A. (1996b) Landscape structure and breeding bird distribution in a sub-Mediterranean agro-ecosystem. *Landscape Ecology* (in press).

Farina, A. (1997) Landscape structure and breeding bird distribution in a sub-Mediterranean agro-ecosystem. *Landscape Ecology* (in press).

Fuhlendorf, S.D. and Smeins, F. (1996) Spatial scale influence on longterm temporal patterns of a semi-arid grassland. *Landscape Ecology* **11**: 107–113.

Gardner, R.H., O'Neill, R.V., Turner, M.G., Dale, V.H. (1989) Quantifying scale dependent effects of animal movement with simple percolation models. *Landscape Ecology* **3**: 217–227.

Greig-Smith, P. (1964) *Quantitative plant ecology*, 2nd edn. Butterworths, London.

Horne, J.K. and Schneider, D.C. (1995) Spatial variance in ecology. *Oikos* **74**: 18–26.

Hornung, M. and Reynolds, B. (1995) The effects of natural and anthropogenic environmental changes on ecosystem processes at the catchment scale. *TREE* **10**: 443–448.

Krummel, J.R., Gardner, R.H., Sugihara, G., O'Neill, R.V., Coleman, P.R. (1987) Landscape patterns in a disturbed environment. *Oikos* **48**: 321–324.

Levin, S.A. (1992) The problem of pattern and scale in ecology. *Ecology* **73**: 1943–1967.

Lubchenco, J., Olson, A.M., Brudbaker, L.B. *et al.* (1991) The sustainable biosphere initiative: an ecological research agenda. *Ecology* **72**: 371–412.

MacArthur, R.H. and Wilson, E.O. (1967) *The theory of island biogeography*. Princeton University Press, Princeton, NJ.

Magnuson, J.J. (1990) Long-term ecological research and the invisible present. *BioScience* **40**: 495–501.

Maguire, D.A. (1985) The effect of sampling scale on the detection of interspecific patterns in a hemlock–hardwood forest herb stratum. *American Midland Naturalist* **113**: 138–145.

Maurer, B.A. (1985) Avian community dynamics in desert grassland: observational scale and hierarchical structure. *Ecological Monographs* **55**: 295–312.

Meentemeyer, V. and Box, E.O. (1987) Scale effects in landscape studies. In: Turner, M.G. (ed.), *Landscape heterogeneity and disturbance*. Springer-Verlag, New York, pp. 15–36.

Morris D.W. (1985) Tests of density-dependent habitat selection in a patchy environment. *Ecological Monographs* **57**: 269–281.

Morris, D.W. (1987) Ecological scale and habitat use. *Ecology* **68**: 362–369.

Mueller-Dumbois, D. and Ellemberg, H. (1974) *Aims and methods of vegetation ecology*. John Wiley and Sons, New York.

Nardelli, R. (1996) *Distribuzione ed abbondanza del pettirosso (Erithacus rubecula) attraverso un gradiente ambientale del sistema appeninico tosco-emiliano*. Unpublished Thesis Dissertation, Parma University, 1995–96.

O'Neill, R.V., Milne, B.T., Turner, M.G., Gardner, R.H. (1988) Resource utilization scales and landscape pattern. *Landscape Ecology* **2**: 63–69.

O'Neill, R.V., Johnson, A.R., King, A.W. (1989) A hierarchical framework for the analysis of scale. *Landscape Ecology* **3**: 193–205.

O'Neill, R.V., Turner, S.J., Cullinan, V.I. *et al.* (1991) Multiple landscape scales: an intersite comparison. *Landscape Ecology* **5**: 137–144.

O'Neill, R.V., Hunsaker, C.T., Timmins, S.P. *et al.* (1996) Scale problems in reporting landscape pattern at the regional scale. *Landscape Ecology* **11**: 169–180.

Powell, T.M. (1989) Physical and biological scales of variability in lakes, estuaries, and the costal ocean. In: Roughgarden, J., May, R.M., Levin, S.A. (eds.), *Perspectives in ecological theory*. Princeton University Press, Princeton, pp. 157–176.

Qi, Y. and Wu, J. (1996) Effects of changing spatial resolution on the results of landscape pattern analysis using spatial autocorrelation indices. *Landscape Ecology* **11**: 39–49.

Steele, J.H. (1989) Discussion: scale and coupling in ecological systems. In: Roughgarden, J., May, R.M., Levin, S. A. (eds.), *Perspectives in ecological theory*. Princeton University Press, Princeton, pp. 177–180.

Stratton, D.A. (1994) Genotype–environment interactions in *Erigeron annuus* show fine-scale selective heterogeneity. *Evolution* **48**: 1607–1618.

Stratton, D.A. (1995) Spatial scale of variation in fitness of *Erigeron annus*. *American Naturalist* **146**: 608–624.

Swanson, F.J. and Sparks, R.E. (1990) Long-term ecological research and the invisible place. *BioScience* **40**: 502–508.

Turner, M.G, Dale, V.H., Gardner, R.H. (1989a) Predicting across scales: theory development and testing. *Landscape Ecology* **3**: 245–252.

Turner, M.G., O'Neill, R.V., Gardner, R.H., Milne, B.T., (1989b) Effects of changing spatial scale on the analysis of landscape pattern. *Landscape Ecology* **3**: 153–162.

Walker, D.A., Halfpenny, J.C., Walker, M.D., Wessman, C.A. (1993) Long-term studies of snow–vegetation interactions. *Bioscience* **43**: 287–301.

Ward, D. and Saltz, D. (1994) Foraging at different spatial scales: dorcas gazelles foraging lilies in the Negev desert. *Ecology* **75**: 48–58.

Weaver, J.C. (1995) Indicator species and scale of observation. *Conservation Biology* **9**: 939–942.

Wiens, J.A. (1986) Spatial scale and temporal variation in studies of shrubsteppe birds. In: Diamond, J. and Case, T.J. (eds.), *Community ecology*. Harper and Row, New York, pp. 154–172.

Wiens, J.A. (1992) Ecology 2000: An essay on future directions in ecology. *Ecological Society of America Bulletin* **73(3)**: 165–170.

Wiens, J.A. and Milne, B.T. (1989) Scaling of landscape in landscape ecology, or, landscape ecology from a beetle's perspective. *Landscape Ecology* **3**: 87–96.

Wiens, J.A., Addicot, J.F., Case, T.J., Diamond, J. (1986) Overview: the importance of spatial and temporal scale in ecological investigations. In: Diamond, J. and Case, T.J. (eds.), *Community ecology*. Harper and Row, New York, pp. 145–153.

Emerging processes in the landscape 4

4.1 INTRODUCTION

This chapter focuses mainly on processes that operate in a landscape across a range of spatiotemporal scales and which in turn influence many landscape patterns.

Among the huge number of processes acting in a landscape we have selected the most obvious, being those on which we have sufficient information. The framework produced by this analysis is neither uniform nor homogeneous, reflecting the gaps in research and the diversity between conceptual and operative approaches, and must be considered a simple introduction only. The strategy that we have utilized in distinguishing these processes is the result of an effort to describe in a simplified and understandable way the emerging components of landscape complexity.

In some cases it becomes hard to distinguish the patterns from the process role, as in the corridors. These may be physical structures (e.g. a road) or may be invisible (e.g. the migratory routes of birds and butterflies).

Disturbance and fragmentation are related processes that in turn determine landscape heterogeneity. They have effects and relationships with many other abiotic and biotic processes, according to different hierarchical levels of function and pattern.

Most of the processes described in this chapter are related to each other. Disturbance and fragmentation are two processes with strong relationships

and it is difficult to distinguish the role and the rate of the interactions. Disturbance has a pre-eminent position in landscape functioning and is the main force responsible for shaping landscapes and their components. This process is driven by many factors and interacts with other processes acting in a more restricted context, such as fragmentation or land abandonment.

According to different perspectives fragmentation can be considered the negative image of connectivity. These processes have a strong influence on the dynamics and fate of material and energy moving across a landscape.

4.2 DISTURBANCE

4.2.1 Introduction

Disturbance is a very common and widespread phenomenon in nature, and may be defined as a discrete event along time that modifies landscapes, ecosystems, community and population structure, changing the substrate, the physical environment and the availability of resources (White and Pickett 1985). It can be considered as a basic process responsible for many other processes, such as fragmentation, animal movements, local and regional extinction etc.

Every landscape is shaped, maintained and/or changed by disturbance. For instance, disturbances such as clearcutting and fire have a strong influence on the structure and functioning of landscapes.

Disturbance is very common in different systems, working at all spatiotemporal scales and altering resource availability and the structure of the system.

Disturbance occurs in many biotic assemblages and at all levels of organization, from individual to landscape. Disturbance combines long timescale changes with reality.

The basic variables of disturbance are magnitude, frequency, size and dispersion. To predict the impact of a disturbance regime on communities and landscape it is necessary to understand at least the spatial and temporal architecture of the disturbance, as discussed by Moloney and Levin (1996).

The mean attributes of any disturbance regime are size, interval and intensity. Their distributions are quite different, as shown in Fig. 4.1.

Disturbance is a source of spatial and temporal heterogeneity. At the landscape level disturbance is related to patch structure and spatial arrangement, and determines the fate of patches and their size and duration (Pickett and White 1985). Severe disturbances or lack of disturbance generally have depressing effects on diversity, but intermediate disturbance seems to enhance diversity in a system.

Where disturbance occurs more frequently than the time required for competitive exclusion, the diversity of the biological assemblage is maintained. This is the case with montane meadows, in which the human disturbance of summer hay harvesting and late summer and early autumn grazing

prevents the development of a dominant species, and as a consequence a higher level of floral diversity is maintained compared to less disturbed prairies. This system appears extremely fragile and needs to be maintained by external input. The fate of such landscapes will be discussed in more detail in the chapter on land abandonment.

Disturbance is the driver of the landscape dynamic and acts at all spatiotemporal scales. Disturbance may be produced by abiotic factors such as solar energy, water, wind, landslides or by biotic elements such as bacteria, viruses, plant and animal competition etc.

All these components are described in other chapters focused on landscape dynamics.

The landscape is differently affected depending on the different disturbance regimes. Gluck and Rempel (1996) have compared the landscape patterns of two forested areas of northwestern Ontario subjected respectively to clearcutting and fire. Patches in the clearcut area were found to be larger in size and with more irregular edges and more core areas than those in the wildfire landscape.

4.2.2 Snow cover, an example of abiotic disturbance

Owing to high climatic constraints the vegetation cover in alpine-type landscapes is extremely patchy. Snow cover controls the distribution of many species of plants, reducing the length of the growing season. The distribution of snow is strongly conditioned by topography and wind patterns, but exposed ridges experience low temperatures in winter and animals such as moles, gophers, voles etc., responsible for the fine-scale mosaic, are also conditioned by snow accumulation, finding refuges and protected trails for soil exploration. The snow vole (*Microtus nivalis*) often builds a nest composed mainly of dry grasses and moss on the surface of the soil, and digs tunnels in the compact snow to search for food (Fig. 4.2).

Plants react to deep snow in different ways (see Walker *et al.* 1993 for more details on the plants of the Rocky Mountains): some species escape deep snow cover (for example *Paronychia pulvinata*), others are mainly localized in deep snow cover (for example *Sibbaldia procumbens*) and others show no precise snow interaction. These plants are good indicators of plant association and can therefore be used as indicators of landscape-scale plant community distribution.

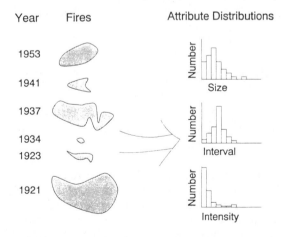

Figure 4.1 Example of fire disturbance regime along a recent period. Every disturbance event (in this case fire) has a disturbance patch with a different shape and attributes (size, interval and intensity) (from Baker 1992, with permission).

Figure 4.2 Snow accumulation during winter in a soil depression creates temporary habitats for micromammals such as the snow vole (*Microtus nivalis*) (M. Cavalbianco, northern Apennines, Italy, 1800 m a.s.l.). The patchy distribution of these accumulations influences vegetation, especially in springtime, owing to the different soil temperature and water content.

Snow accumulation has indirect effects on vegetation and on the circulation of nutrients in the soil during the spring thaw. This process is extremely important in alpine regions where plants suffer from nitrogen and phosphorus deficits (Fig. 4.3).

4.2.3 Human disturbance

Human disturbance is really no different from natural disturbance, but there are some significant differences especially in extension, severity and frequency (Fig. 4.4). Forestry, agriculture, development and infrastructure are some of the disturbances that human activity can produce on the landscape and, at a larger scale, on regions.

Human–environment interactions are distributed worldwide and the whole planet is affected. Naveh (1992) uses the term 'total human ecosystem' to refer to the earth. Disturbance due to human activity is expanded by technology to a broad range of spatiotemporal scales, with effects spreading from the oceans to the highest mountain ridges. The capacity of the landscape to incorporate human disturbance is in many cases overwhelmed and the disturbance process is transformed into a stress process that reduces diversity.

Often the attributes of human disturbance differ from those of a natural disturbance. For instance, a fire along a Mediterranean coast produced intentionally or by human carelessness is no different as a process from wildfire, but if repeated at every season (improbable in wild conditions) can cause stress to vegetation, reducing vegetation cover and increasing soil erosion.

4.2.4 Gap disturbance in forests

Gaps are small openings in forest cover caused by local events such as tree fall, and generally are not a random event, but some sites are more likely to have gaps than others (Poorter *et al.* 1994). Most of the species that live in old-growth forests require such gaps to reach maturity. Across the forest the gap density is constant, but these authors have found that along a catena, gaps are more abundant in the middle part, whereas the upper and lower slope gaps are less frequent.

The regeneration that occurs after the formation of a new gap plays a fundamental role in structuring forests and maintaining species diversity. Gaps are inhabited by species of organisms different from those in the forest understorey. Large forest disturbances such as hurricanes can change the structure of bird assemblages in old gaps, and cause new forest disturbance. Wunderle (1995) found that the main effect of the passage of Hurricane Hugo on bird assemblages in a Puerto Rican forest was the loss of distinctiveness between those living in the gaps and birds living in the disturbed understorey. According to this author probably many years will pass before gap and understorey once more become distinct in structure and resources.

In forests in which large disturbances are rare, the gaps created by the killing of one or a few canopy trees play a fundamental role in structuring the entire forest. Gaps are particularly evident in the changing phase from mature to old-growth forests. A tree that dies is considered a gapmaker because it creates a gap. An edaphic gap is distinguished from a tree gap because it is produced by edaphic conditions such as a watercourse or thin soil.

In forests of British Columbia Lertzman *et al.* (1996) have estimated that in the absence of large disturbances such as fire, windstorms or insect diseases, gaps created by a regime of small-scale, low-intensity disturbance are responsible for the turnover of these forests in between 350 and 950 years. Gap disturbance is common, involving about 56% of the forest area investigated. Most of the gaps are produced by the death of more than one tree. Some gaps (a third) were found to be caused

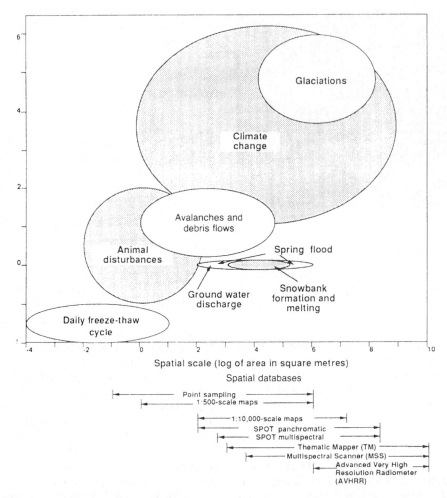

Figure 4.3 Frequency of the disturbance and the spatial scale of resolution in a cold climate. The available data types are indicated at the bottom of the figure as an example of the application of a multiscale approach, ranging from data input by field survey (quadrant plots) to remote sensing technique (advanced very high resolution radiometer, AVHRR) (from Walker *et al.* 1993, with permission).

by edaphic factors. Figure 4.5 shows the frequency distribution of edaphic gaps.

4.2.5 Gaps in savanna

Recently Belsky and Canham (1994) have discussed the structure and function of savanna trees in a matrix of grasslands, comparing forest gaps with savanna trees. This approach seems very interesting, although in some cases intriguing. Tropical savanna dynamics may be explained in terms of gap dynamics: in the grassland matrix trees and shrubs

are the 'gaps'. Figure 4.6 shows an example of forest and savanna gaps.

In the savanna a gap is created by tree seedling establishment and growth. As with gaps in the forest environment the physical conditions under savanna trees are different from those in the surrounding open savanna. During wet seasons the soil under trees is dryer, but later in the season it is wetter because of reduced evapotranspiration in the shade and a cooler temperature.

It is well known that under savanna trees the soil is richer in nutrients, owing to root transportation,

Figure 4.4 Human disturbance may be caused by different types of activities. (**a**) Mining (Alpi Apuane marble mines); (**b**) by recreation infrastructures (sky facilities) (northern Apennines); (**c**) by restoration plans not ecologically oriented (ploughing in an arid–rocky mountain landscape in southern Spain, to improve the pine plantation); (**d**) gravel mining in an active river bed (Aulella river, northern Apennines); (**e**) coppicing, matricinato selective cut (northern Apennines); (**f**) prescribed fire in sweet chestnut orchard (northern Apennines). All these activities, because of the severity of the disturbance or the fragility of the ecosystem, cannot be incorporated in the landscape.

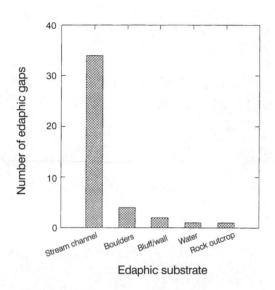

Figure 4.5 Frequency distribution of edaphic gaps distinguished by physiographic causes (from Lertzman *et al.* 1996, with permission).

manure deposition by both wild and domestic animals, and a less stressful bacteria cycle.

4.2.6 Fire disturbance in dry landscapes

Fire is one of the most important shaping agents in dry landscapes. It removes the undecomposed biomass and creates nutrient fluxes by ash deposit watering, contributing to the ecologically rejuvenating qualities of forest ecosystems (Moore 1996).

Fire has been utilized as a management tool to manipulate the ecosystem since Mesolithic times (Naveh 1990, 1991). However, if the release of nutrients is well documented, the role of charcoal in the soil appears neglected. Charcoal has the capacity to retain water, and a sandy soil can behave like clay if conditioned by charcoal. However, charcoal has recently been recognized as a sink for phenolic inhibitors, providing stimulation for both plants and microbes.

4.2.7 Pathogen disturbance

Little attention has been given to the role of pathogens in shaping and structuring forests. Pathogens influence forests at different ranges of

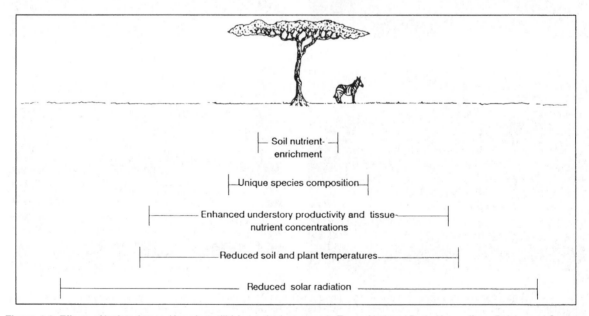

Figure 4.6 Effects of isolated trees (*Acacia tortilis*) in tropical savanna in Tsavo National Park, Kenya (from Belsky and Canham 1994, with permission).

spatial and temporal scale. Some, like *Brunchorstia pinea*, responsible for the Scleroderris canker in pines, are more active in cold temperatures, having a greater impact on stands occurring in topographic depressions and forest openings where cold air accumulates.

In combination with land abandonment, the canker of sweet chestnut has accelerated change in the landscape of southern Europe. A large area occupied by this type of orchard has in a short time been modified by cutting of diseased plants.

Pathogens play a fundamental role in the formation of gaps in mature and rich old-growth forests. Patch-phase processes of disturbance create the conditions for landscape heterogeneity, enhancing plant diversity and the resources available for both plants and animals. Pathogens can also change the composition of forests, increasing the unevenness of stands. A knowledge of pathogen cycles is essential for efficient and accurate management of forests (Castello *et al.* 1995).

4.2.8 Animal disturbance

Grazing is the most common disturbance caused by animals (herbivores) (Fig. 4.7). This disturbance severely affects the distribution and structure of vegetation and in forested areas can prevent the growth of seeds. Trampling associated with grazing modifies the composition of natural vegetation and reduces interspecific competition, creating patches of high diversity (Fig. 4.7). The disturbance regime is often quite complex (Figs 4.7–4.10).

Figure 4.7 Terracettes associated with domestic sheep grazing in the M. Sillara, northern Apennines (Italy).

Figure 4.8 Ants' nests in montane moorlands create disturbance patches, increasing the spatial heterogeneity.

Figure 4.9 The mole (*Talpa* sp.) is a disturbance agent in open and forested landscape. Mound deposits are centres for seeding of non-dominant plants.

In both natural and livestock-grazed prairies the deposition of urine is a cause of local disturbance that at a larger scale produces a complicated mosaic. This mosaic depends mainly on the density of grazing wild or domestic animals. Steinauer and Collins (1995) tested the effect of urine deposition on differently disturbed grasslands. Plant abundance increased after urine deposition, but alpha and beta diversity had local behaviour, mainly due to litter depth. This biomass accumulation attracts more herbivores and the effect of urine is expanded into the neighbouring environment. Finally, grazing intensity in such patches has a more profound effect

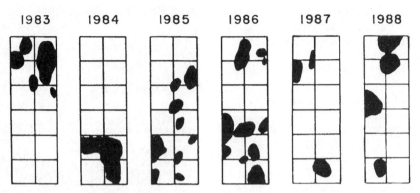

Figure 4.10 Spatial distribution of gopher disturbance by digging in a plot of 3×1 m from 1983 to 1988 in a serpentine grassland (from Hobbs and Mooney 1991, with permission).

than urine deposition alone. This seems a good example of disturbance overlap, which can reinforce the reaction of the environment.

4.2.9 Summary

- Disturbance is a very common process in most ecological systems, acting at different spatial and temporal scales.
- Size, interval and intensity are the main attributes of a disturbance regime.
- Abiotic disturbances such as snow cover are responsible for the distribution and abundance of many species of organisms.
- Human activity is one of the most diffuse disturbance regimes, affecting many ecological entities from species to whole communities and ecosystems.
- Gaps in forests caused by tree fall, or in savannas by tree growth, represent a common source of disturbance. In old-growth forests gaps due to tree fall are fundamental to long-term forest dynamics.
- Animals are important causes of natural disturbance. Trampling, grazing and nitrogen deposition modify the landscape, enhancing spatial and temporal heterogeneity.

4.3 FRAGMENTATION

4.3.1 Introduction

Several papers have recently focused on the fragmentation process as a central issue in landscape ecology and conservation planning (e.g. Wilcove *et al.* 1986, Saunders *et al.* 1991). Fragmentation is one of the most severe processes to depress biodiversity. It moves at an alarming rate around the world, reducing large forest cover as well as natural prairies. In some parts of the earth fragmentation occurred mainly in the last century, for example in Australia and Brazil (Hobbs and Hopkins 1990), with devastating consequences on the environment (Fig. 4.11).

Fragmentation has a negative influence on many species of plants and animals, and on ecological processes. The smaller the fragmented blocks the more the density of populations decreases and the risk of extinction grows. Fragmentation means geographical isolation, and after extinction the probability of recolonization strongly depends on the distance of fragments from the main core and on the quality of the surrounding habitat.

Edge-sensitive species (interior species) can reduce in abundance and large predators, for instance, disappear, producing outbreaks of foragers such as deer. This last effect is a key starting point for further environmental degradation and disturbance.

Fragmentation is often interpreted by the general framework of the island biogeography theory (MacArthur and Wilson 1967) but area size and isolation factors taken into account by this theory are not enough to explain the effect of fragmentation in habitat islands. If fragmentation is considered simply as the size of isolated patches, this approach appears uninformative. Fragments cannot be considered true islands: in fact, the surrounding habitat often is not completely hostile to the species.

Other factors, such as connectivity, the presence of ecotones and corridors, and the metapopulation structure have to be taken into account, especially when fragmentation is studied at a landscape scale.

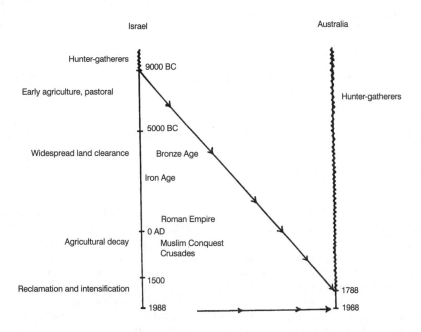

Figure 4.11 A comparison of the main modifications of natural vegetation over centuries in Israel and Australia (from Hobbs and Hopkins 1990, with permission).

Fragmentation can be considered a continuum process and, according to a landscape perspective, matrix and patches are the elements that have to be used when considering a landscape, fragmented or not (Wiens 1994) (Fig. 4.12).

Fragmentation is perceived in different ways by different species, and even for the same species it may be different according to the season.

At the edge the behaviour of a species is different. For some species edges are highly suitable habitats, but others avoid them. Nest predation can be higher at the edges, and this has a big influence on the suitability of the patches in a fragmented landscape.

Fragmentation is really a dynamic process. Human as well as natural disturbance produces fragmentation, but often the recovery of vegetation cover masks or mitigates the process. In other cases the fidelity of a species to a site reduces the effect of fragmentation.

4.3.2 Scale and patterns of fragmentation

Fragmentation is a scale-dependent process. Fragmented vegetation can have a different spatial arrangement, causing different effects on other ecological processes.

To describe the dispersion of fragments in an area it is necessary to consider their different attributes, such as density, isolation, size, shape, aggregation and boundary characteristics.

The isolation of patches increases geometrically as the density of fragments decreases. The smaller the fragments the more they are influenced by the surrounding matrix. If fragments are aggregated their isolation is less than in conditions of equi-dispersion.

There are different types of fragmentation. According to Lord and Norton (1990), when an intact area is divided into smaller intact fragments we have a 'geographical fragmentation'. This process has received a lot of attention from conservation ecologists because of its implications for nature conservation. This pattern is analogous to a coarse-grained landscape (Table 4.1).

At the other extreme we can have fragments at the scale of individuals or small plots. Where small remnants of native vegetation are embedded in an alien matrix, the fragmentation is considered to be 'structured' and is analogous to a fine-grained landscape. Fine-grained fragmentation generally pre-

Table 4.1 Implication of different scales of habitat dispersion to various attributes of fragments and fragmented landscapes (From Lord and Norton 1990, with permission)

	Geographical	*Structural*
Size (m²)	large >1000	small <10
Isolation	usually medium to large	usually small
Boundary gradient	steep	shallow
Impact of extrinsic disturbance	confined to edge and up to a few hundred metres	throughout
Vulnerability to functional disruption	medium to small	medium to large
Scale of organism affected	large generalist to medium specialist	medium specialist to small specialist
Advantages for conservation	usually has intact interior	usually greater total extent

sents patches close to each other and the contrast between patches and matrix is shallow, creating a pseudocontinuum.

Whereas geographical fragmentation is associated with forest ecosystems, structural fragmentation may be associated with a broad range of conditions.

The effects of fragmentation on organisms depends largely on the scale of species-specific perception: generalists are less affected by fine-grained fragmentation than are specialists.

Fragmentation increases the vulnerability of patches to external disturbance, for instance windstorm or drought, with consequences for the survival of these patches and of the supporting biodiversity (Nilsson and Grelsson 1995).

The scale of fragmentation has a direct impact on organisms. Large fragments maintain a good subset of species, but small fragments have only few species, usually the more generalists. So, specialists disappear from smaller patches when fragmentation is at a fine scale. This could be the reason for diversity in a temperate region facing fine-grained fragmentation, compared to the more specialist species of tropical areas.

The patterns of fragmentation are affected by many natural and manmade variables. Agricultural proximity is a good indicator of fragment probability in the bottomland of a hardwood forest, but access, urban development, ownership, fencing and regional differences are other, secondary, parame-

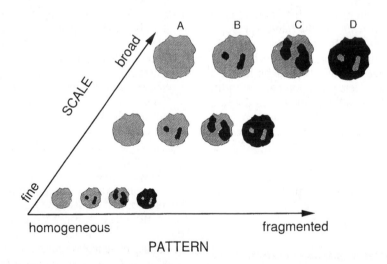

Figure 4.12 Pattern and scale from homogeneous to fragmented. Fragmentation occurs at every scale and appears as a continuum. A landscape can be considered fragmented at one scale and homogeneous at another (from Wiens 1994, with permission).

ters useful for predicting the type and modality of fragmentation (Rudis 1995).

Fragmentation reduces the size of woodlot but also the habitat quality. Large fragments have more species, are less disturbed and have lower road access than smaller fragments. Large fragments are uncommon or rare, and their importance is great for nature conservation issues.

Fragmentation depends on human use, but human use is also affected by the fragmentation rate. This is relevant in regions such as the Mediterranean, where several changes in land use have occurred over the centuries, modifying the behaviour of people according to the new characters of fragments.

Fragmentation can be observed at any scale, and tree-fall gaps are one important factor increasing heterogeneity and fragmentation. Tree-fall gaps can be considered a distinct habitat in a forest, differing in vegetation structure (e.g. foliage density, tree size distribution), plant species composition, resource abundance and microclimate conditions. Birds are particularly sensitive to slight changes in habitat features and probably have the capacity to recognize tree-fall gaps, especially as preferred foraging sites. In fact, in tree-fall gaps resources are particularly abundant, depending on the gap size, the age of the gaps and the surrounding forest track. In spring many migratory birds are attracted to these gaps because food resources are more abundant (Blake 1986, Blake and Hoppes 1986).

4.3.3 Community composition and diversity in fragments

Small woodlots have fewer species than large ones, and there are more generalists. The number of more specialized species increases with the increase in woodlot area. Blake and Karr (1987) found that more than 66–72% of species are strongly influenced by habitat variables. For example, birds breeding in the forest interior and wintering in the tropics are more affected by fragmentation of their habitat (Fig. 4.13).

The area effect is disputable according to neighbouring habitats. In fact, if there are suitable habitats around a woodlot these could be incorporated by some species, but if the woodlot is separated by agricultural fields the habitat constraint is much stronger and the isolation greater, negatively affecting the presence of species.

The continuous loss of forests across the USA will probably have a negative effect on birds,

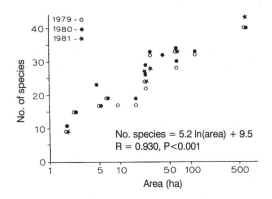

Figure 4.13 Number of breeding species plotted as a function of natural logarithm of area of woodlot in east-central Illinois (from Blake and Karr 1987, with permission).

although a three-year study has demonstrated good stability in woodlot populations (Blake and Karr 1987). However, this does not mean that over a longer period negative effects will not appear, and consequently monitoring is needed to control these trends.

Large patches of nothofagus forest in southern temperate rainforest are more heterogeneous than smaller patches. Bird diversity decreases according to patch size. This effect is also evident where patches are not far from the main forest or from other patches, and where apparently shrubby corridors occur. For instance, the main causes of the decrease in abundance and diversity of Chilean avifauna depend on habitat destruction. Clearing of understorey may also be a negative factor, because many species breed in the shrub layer or find resources at this height (Willson *et al.* 1994).

When tropical forests are fragmented an immediate loss of biodiversity is experienced. Despite the temperate fragmented forests in which diversity, especially of birds, is maintained relatively high and species travel long distances to colonize a site, in the tropical forests short distances between fragments and the continuous forest are a real barrier to the movements of birds. Insectivorous birds are heavily affected by isolation: obligate army ant followers disappear within 2–3 years of isolation.

On isolation after logging or agricultural use a fundamental role is played by the surrounding vegetation. Woodlots surrounded by *Vismia*, a dominant vegetation after forest removal by burning and cattle pasture, are considered more isolated by birds than plots surrounded by *Cecropia*, a vegetation that occurs where forest is removed by logging but not by burns (Stouffer and Bierregaard 1995).

Tropical rainforests cover less than 7% of the planet's land mass but support half to two-thirds of its plants and animals. The sensitivity of tropical forests to fragmentation has been investigated by Bierregaard *et al.* (1992) as part of the Biological Dynamics of Forest Fragments Project. Distance effects, fragment size, edge effects and biotic changes were some of the more important issues; 80 m of non-forest is enough to create a barrier for mammals, insects and understorey birds living in a fragment.

The sizes of fragments at which most species of insects, mammals and birds become sensitive are 1, 10 and 100 hectares. After fragmentation birds move across fragments more, and density increases at least 200 days after logging. This clearly show how birds have the capacity to move away from fragments.

The hypothesis that small patches contain subsets of organisms of larger ones is discussed by Cutler (1991) (Fig. 4.14), studying mammals in the Great Basin of western North America. Actual mammal composition is the result of selective, deterministic extinction of species of originally richer fauna. In the Florida Keys deforestation is producing a decrease in bird numbers. For some sensitive species the loss of habitat is more important than the actual loss of deciduous forest. This means that habitat requirements for sensitive species are not limited to area size, but also to surrounding characters (Bancroft *et al.* 1995).

Fragmentation in shrub steppe habitats negatively affects the breeding distribution of shrub-obligate species (Knick and Rotenberry 1995). The difficulties of shrub steppe restoration may cause irreversible loss of habitat, and negative long-term consequences for shrub-obligate species.

In old-growth montane forests on Vancouver island (Pacific Canada) the effect of fragmentation of bird assemblages was less dramatic than in other areas (Schiek *et al.* 1994), probably because the old-growth forest evolved within heterogeneous montane forests. It could also be because of the lower contrast between old-growth and logged areas, compared to forests and agricultural/urban areas where most fragmentation studies were carried out.

Tscharntke (1992) has found that fragmentation of *Phragmites* habitats is having a profound effect on insects and birds. An important point stressed by

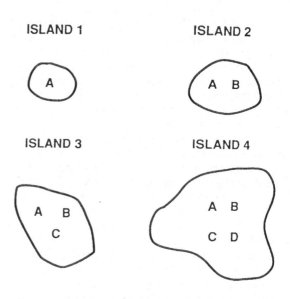

Figure 4.14 The species found in fragments are subsets of assemblages of larger plots (from Cutler 1991, with permission).

this author concerns the significance of habitat fragmentation that is limited not only to the size of patches, but also to the mean shoot diameter. Further, small patches of *Phragmites* receiving more light have greener leaves than individuals in large dense patches, and some species of aphids are positively correlated with fragmentation.

The fragmentation of midwestern grasslands has impoverished the breeding bird communities (Herkert 1994). Five species (savannah sparrow, Henslow's sparrow, bobolink, eastern meadowlark) have been found sensitive to patch size. The increase of farmland has dramatically reduced the size and distribution of native grasslands, a fact which has determined the decline of many bird species (Fig. 4.15).

4.3.4 Species, guilds and fragmentation

Some species are sensitive to habitat size and are called 'area-sensitive'. Thus forest-interior breeding species such as the oven bird and Kentucky warblers are in decline owing to fragmentation (Gibbs and Faarborg 1990).

The effect of fragmentation of tropical forest on dung and carrion beetle communities was studied by Klein (1989) in Central Amazonia. Fragments had fewer, more rare and more dispersed species. This

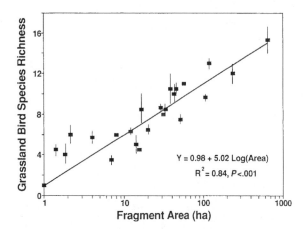

Figure 4.15 Bird species richness of grasslands plotted against the area of fragments. Symbols represent number of species found per site in 1987–1989 censuses (from Herkert 1994, with permission).

difference was particularly evident when a 1 hectare fragment was compared with undisturbed forest. The movement of beetles was interrupted by clearcutting, and only after a secondary regrowth were some beetles able to move in fragments, although the distance was only a few metres. Probably the microclimate, and especially the desiccation at the border of the woodlot, creates prohibitive conditions for the forest understorey beetles.

The role of dung and carrion beetle as destroyers of nematode larvae and other gastrointestinal parasites of vertebrates is very important in disease control. With increasing fragmentation of tropical forests and the consequent reduction in the numbers of such beetles, we can expect an increase in diseases of vertebrates.

When the effects of habitat fragmentation are observed at the scale of singular species or groups of related species it is possible to find interesting surprises and unpredicted results. An example is the scorpion *Cercophonius squama* and the amphipods, family Tallitridae, as reported by Margules *et al.* (1994). In a time lag of 8 years, 3 before fragmentation and 5 after fragmentation in treatment and control plots of Australian hardwood forest dominated by *Eucalyptus*, the fragmentation had no effect on the abundance of scorpions. Results were different for the amphipods, which decreased markedly after fragmentation, more in small than in larger remnants (Figs 4.16 and 4.17).

Probably the scorpion, an ancestral animal, using a fossorial behaviour during dry periods, has the capacity to escape the environmental stress of fragmentation. This is not the case with amphipods, which are more sensitive to microclimate and have a younger evolutionary history. This is a good illustration that when we study fragmentation at the level of the species the results may be completely different.

Management should be carried out according to the behaviour of populations and landscape scale and subpopulation in the remnants, rather than considering simply the number of species. *Malacosoma disstria* is a forest tent caterpillar that has outbreaks of different duration according to forest structure (Roland 1993). The amount of forest edge per square kilometre is a good predictor of the duration of outbreaks. Parasitoides are probably less efficient in controlling *Malacosoma disstria* along fragments. As with many lepidopteran species, *Malacosoma disstria* lays more eggs along the edges and egg predation on this species is apparently at a lower level than in the forest interior.

Along edges the more favourable microclimatic conditions and higher temperatures decrease the period of larval development, thereby reducing the risk of predation compared to larvae in continuous stands (Fig. 4.18).

Species sensitive to habitat fragmentation are less efficient at moving and colonizing new habitats, and consequently have a lower dispersal ability (Villard and Taylor 1994, Villard *et al.* 1995). Birds of the African tropical forest understorey are particularly sensitive to fragmentation (Newmark 1990). Rare and forest-interior species are the ones most adversely affected by habitat reduction, and the maintenance of corridors could be a local extinction-mitigation action.

Insect pollinators in the subtropical dry forests of northern Argentina declined, decreasing the size of fragments, but conversely the honey bee (*Apis mellifera*) increased its frequency of flower visits (Aizen and Feinsinger 1994).

Despite an expected high rate of alien species in small isolated woodlots, invaders are stopped by the light concurrence (low in interior woodlots) and by cropland isolation, combined with the low capacity to move of alien species. The fragments have edges of dense shrubby vegetation that prevents alien species from entering, although at the same time these warm edges are attractive to alien species (Brothers and Spingarn 1992).

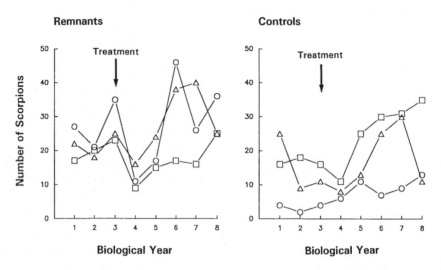

Figure 4.16 Total numbers of the scorpion *Cercophonius squama* recorded in each forest remnant size in each year, open circle (small), square (medium), triangle (large) (from Margules *et al.* 1994, with permission).

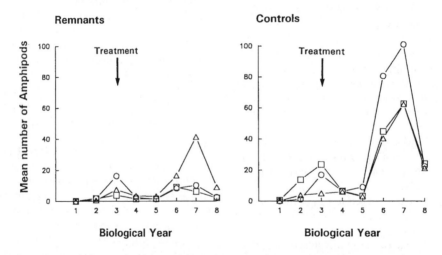

Figure 4.17 Mean numbers of the amphipod, family Tallitridae, in each forest remnant size in each year, open circle (small), square (medium), triangle (large) (from Margules *et al.* 1994, with permission).

4.3.5 Habitat fragmentation and extinction

The effect of fragmentation on bird species diversity has been documented in the San Antonio upland forest of Colombia by Kattan *et al.* (1994). In this area the avian census of 1911 was compared with data from 1959, 1963 and 1989–90. The loss of 24 species, or 31% of the original avifauna, is relevant information on the level of fauna impoverishment

in these landscapes. The authors who carried out the more recent census and the comparison argued that the high rate of extinction is mainly due to the position of many species at the upper limit of their altitudinal distribution, and that the more vulnerable assemblages were understorey insectivores and large canopy frugivores.

The work of Kattan *et al.* (1994) is important from many points of view, first because the history

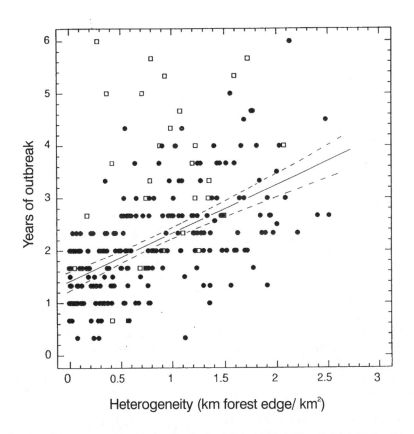

Figure 4.18 Mean duration of forest tent caterpillar outbreaks for 261 townships (from Roland 1993, with permission).

of bird decline is fully documented, and secondly because the effect of fragmentation depends in large part on the biogeography of the species and on the complexity of foraging structure.

4.3.6 Predation and fragmentation

Isolated woodlots generally experience more predation. Wilcove (1985) found a higher rate of predation in small than in large woodlots, but predation was also higher in suburban neighbourhoods than in isolated farmland woodlots. Placing artificial open-cup nests on the ground and at 1–2 m above ground, predation was found to be higher in both cases than in experimental cavity nests. Considering that most of the neotropical migratory songbirds construct open-cup nests, the decline of these species due to predation can be more than just suspicion (Fig. 4.19).

In large forest blocks in Virginia predation on artificial nests varied from 5 to 40%, according to many vegetation variables and predator community pressure (Leimgruber *et al.* 1994).

Fragmentation of prairies and marshlands is moving very quickly in some regions of the world, such as Canada, under the impact of modern industrialized agriculture. Although many aquatic birds have resilient capacities to buffer the habitat loss, the vanishing of large natural breeding habitats tremendously increases the risk of predation. Pasitschniak and Messier (1995), using artificial nests placed at different distances, simulated predation risks at the edges. The predation risk of waterfowl nests is related to distance from the edge in dense nesting cover, but no edge effect has been observed in idle pastures or delayed hayfields. This could be due to the abrupt edge between these cultivations and the greater accessibility for predators.

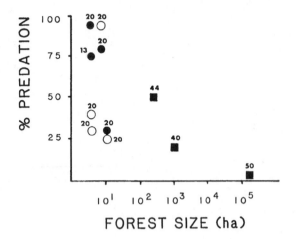

Figure 4.19 Forest size and predation rate. ■ = Large forest tracts, O = rural fragments, ● = suburban fragments (from Wilcove 1985, with permission).

So, the edges in a manmade landscape could be less important than vegetation structure.

This argument has been questioned by other authors. The predation of ground nests in prairie fragments in Missouri was studied by Burger *et al.* (1994). Artificial nests in prairies <15 ha were predated more than those in large prairie remnants (37% vs. 13.9%). Nests placed at a distance <60 m from woodlands were less likely to be successful than artificial nests placed further away (28.7% vs. 7.9% of predation).

The fragmentation of holarctic forests has altered many dynamics of small herbivores compared to those in undisturbed forest. The disappearance of cycles in abundance moving south is mainly due to increased pressure from predators. Andren *et al.* (1985) studied predator pressure on tetraonids using dummy nests. The predation pressure was higher in the south, as predicted, and the main predators are corvids, which are more abundant in the southern regions. These birds are positively correlated to the farming system, to the fragmentation of the forest and to a greater human disturbance regime (Fig. 4.20).

4.3.7 Island size and isolation, a key to understanding fragmentation

It is extremely important to verify the effect of area and isolation in a true island landscape to understand the fragmentation process better.

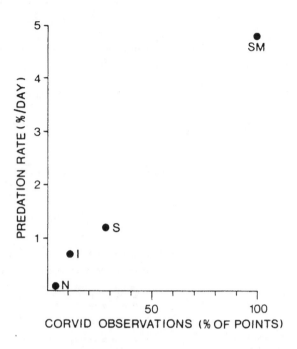

Figure 4.20 Relationship between predation rate and the number of corvid observations in Sweden. N = the northern area, I = the intermediate, S = the southern area, SM = the southernmost area used by Goransson and Loman (1976) (from Andren *et al.* 1985, with permission).

The investigation of Martin *et al.* (1995) in the Gwaii Haanas archipelago of British Columbia on birds breeding in islands from 1 to >100 000 ha in area has demonstrated that many factors have to be considered when the fragmentation process is investigated.

The analysis of the landscape pattern seems extremely important. Four major factors were recognized as being responsible for differences in habitat structure among islands. First, in small islands more edges increase the shrub cover, and the proportion of edge is consequently higher in small than in larger islands. Secondly, small islands are more exposed to windstorms, tree cover is more sparse and there is more undergrowth. Thirdly, the less isolated islands receive more rain than isolated ones, and the moss biomass is consistently smaller. Fourthly, in larger islands the black-tailed deer *Odocoileus hemionus* is more abundant than in smaller islands. In some cases the smallest islands have no capacity to support a permanent deer population. The browsing of deer has a very strong effect on understorey vegetation.

The effect of area on bird species is important only in small islands. With decreasing island size some species became rare but others became more common. A correlation was found between habitat features and birds. Only for a small proportion of species did island size and isolation appear important.

Area *per se* appeared insufficient to explain bird species diversity and abundance in Gwaii Haanas Archipelago, but habitat features, although in some degree related to island size and isolation, are important factors.

4.3.8 Habitat fragmentation and animal behaviour

Habitat fragmentation modifies some aspects of animal behaviour, such as movement and food searching. The response of small mammals to fragmentation was tested by Diffendorfer *et al.* (1995) in three 0.5 ha plots with larger patches 5000 m², medium patches 288 m² and small patches 32 m² from 1984 to 1992; the species studied were the cotton rat (*Sigmodon hispidus*), deer mice (*Peromyscus maniculatis*) and prairie voles (*Microtus ochrogaster*). As expected, animals moved over larger distances, and a lower proportion of animals moved, thereby increasing fragmentation. No source–sink mechanisms were verified.

The tawny owl (*Strix aluco*), a common nocturnal predator in Europe, seems sensitive to woodland fragmentation. The foraging behaviour of 24 owls radio-tracked by Redpath (1995) was analysed both in continuous woodland and in farmland with sparse woodlots. In fragmented woodlots the owls had longer interperch distances and perch times. Males had longer perch times than females, and males in fragmented woodland spend 40% more time in flight than males in continuous woodland. Fragmentation therefore greatly influences owl activity and behaviour (Fig. 4.21).

In some species adaptation to fragmentation can be innate, as is the case with the desert-dwelling mountain sheep (*Ovis canadensis*). This species is adapted to live in steep, mountainous open terrain, naturally fragmented, and has developed a high vagility in fragmented habitats in conditions where suitable corridors can persist (Bleich *et al.* 1990).

Dispersion reduces the isolation effect but encourages the spread of diseases and the greater mortality of moving organisms (Burkey 1995).

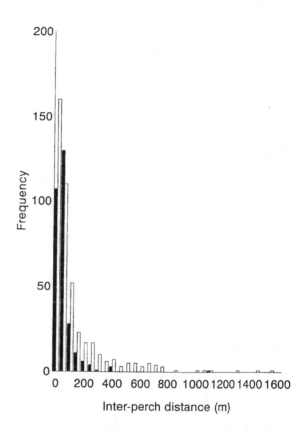

Figure 4.21 Frequency distributions of distances moved by tawny owl (*Strix aluco*) in Monks wood (shaded) and the farmland (open) (from Redpath 1995, with permission).

4.3.9 Measuring the effects of fragmentation

Hinsley *et al.* (1995) have employed many variables to characterize fragmentation in relation to birds (Table 4.2): area, structure, isolation, connectedness and surrounding land use. The turnover of species was calculated according to Diamond (1969):

$$(E + C)/(S1 + S2) \times 100\%$$

where E and C are the numbers of species extinctions and colonizations respectively. S1 and S2 are the number of breeding species present in the two years. Turnover rate of birds was higher in small than large woodlots, although some census biases can cloud the results.

Turnover was found to correlate with plot area, but the higher turnover in small woodlots was due to the change in common species. On increasing the

Table 4.2 Variables describing woodland area and structure, isolation, connectedness and surrounding land use, used in stepwise multiple regression analysis (from Hinsley *et al.* 1995, with permission)

Area and structure

Woodland area (ha)
Perimeter (m)
Shape
Density of canopy layer
Density of shrub layer
Density of field layer
Number of habitats

Isolation

Area of woodland within 0.5 km (ha)
Area of woodland within 1.0 km (ha)
Length of hedgerow within 0.5 km (km)
Length of hedgerow within 1.0 km (km)
Distance to nearest wood (km)
Distance to nearest wood ≥2 ha (km)
Distance to nearest village/town (km)

Connectedness

Number of hedges connected to wood
Number of ditches connected to wood
Number of dirt roads to/from wood
Total number of linear connections

Surrounding land use

% of perimeter adjoined by cereal
% of perimeter adjoined by rape and beans
% of perimeter adjoined by root crops
% of perimeter adjoined by other crops
% of perimeter adjoined by grass
% of perimeter adjoined by non-crop usage

dimensions of the woodlot the turnover remained steady, especially in rare species such as the great-spotted woodpecker, whereas common species were excluded by stochastic extinction.

4.3.10 Summary

- Fragmentation is a worldwide process that reduces biodiversity and accelerates the local and global extinction of plants and animals.

- Habitat fragmentation is species specific, so an environment perceived as fragmented by one species can be perceived as homogeneous by another.
- Fragmentation increases habitat edges and also the risk of predation, considering that many predators prefer edges as a hunting area.
- Fragmentation reduces specialist habitat-demanding species and favours generalists.
- Fragmentation increases patch fragility.
- Tropical species are more vulnerable to fragmentation than temperate ones.
- Animal dispersion and movements increase with fragmentation rate.

4.4 CONNECTIVITY, CONNECTEDNESS AND CORRIDORS

4.4.1 Introduction

Landscapes are heterogeneous across a broad range of scales. This heterogeneity, as discussed in Chapter 5, is determined by the presence of patches with different isolation from patches of the same type. It is obvious that isolation creates problems for the diffusion of organisms and reduces survival capacity when the organisms are few in number. Three important concepts can address the problem of patch isolation: connectedness, connectivity and corridors.

Connectedness is the degree of physical connection between patches. It is a structural attribute of a landscape and can be mapped (Baudry 1984) (Fig. 4.22). Hence the matrix is the most connected element of a landscape. Generally we use connectedness to refer to other elements of the landscape, such as woodlot, hedgerows, riverbeds etc.

In some rural areas the hedgerow network is the most connected component after the field matrix. Woodland connectedness plays a fundamental role for species that need tree cover for their movements. For instance, the increase in woodland connectedness after land abandonment has favoured the diffusion of the wild boar in most of the mountain landscapes of Europe.

Merriam (1984) utilized the term connectivity as a 'parameter to measure the processes by which the

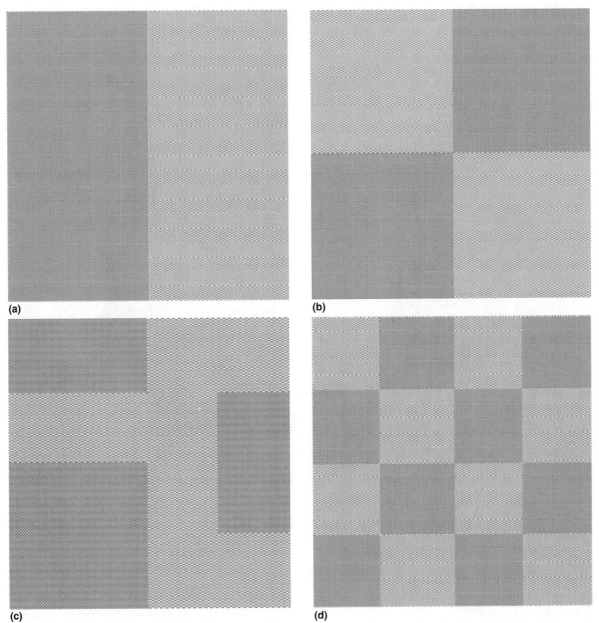

(a)

(b)

(c)

(d)

Figure 4.22 Example of four landscapes with the same percentages of two land covers but spatially arranged according to different values of decreasing connectedness.

subpopulations of a landscape are interconnected into a demographic functional unit'. Connectivity can be seen as the inverse correlate of hostility of interpatch habitat (Fig. 4.23).

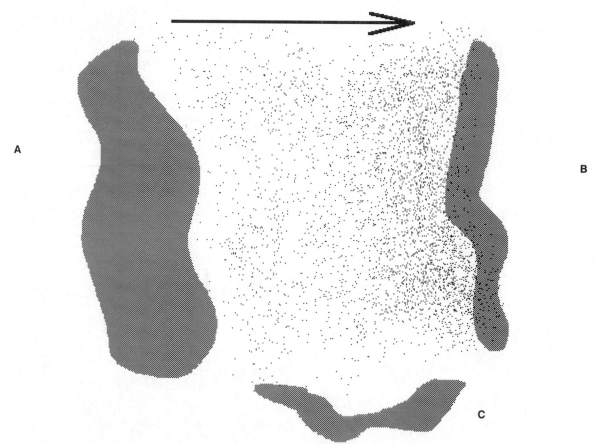

Figure 4.23 Connectivity is a functional parameter that varies according to specific organisms, and is frequently not related to landscape structure. In this case the connectivity between patch A and B is high, whereas patch C (closer to patch A) is more isolated for wind-dispersed organisms if the wind is blowing in the direction indicated by arrows (from Baudry and Merriam 1987, modified).

Landscapes with high connectivity can assure a greater survival probability to isolated populations, as confirmed by Merriam (1984) in the Canadian rural landscape for some species of rodents.

In some cases connectedness is low but connectivity is high, and in this case we can assume that functional corridors exist.

4.4.2 Corridors: structure and functions

Corridors are functional structures in a landscape and their presence is fundamental to mitigate the effects of fragmentation or, vice versa, to increase the penetration of alien species. In other cases corridors are structurally recognizable, such as hedgerows. The corridor concept is not clear and is often used with different meanings. The controversy on the exact role

of corridors in a landscape is open and largely depends on the different contexts in which they are considered (see for example, Simberloff *et al.* 1992).

A corridor can be defined as a narrow strip of habitat surrounded by habitats of other types. Plants and animals can move easily across a corridor, but a great variability in species behaviour exists along corridors.

There is little evidence that animals use structured corridors such as hedgerows and fences. The same is true for many plants that for dispersion, germination and growth need soil conditions that cannot be assured by a narrow belt of vegetation (the corridor).

Corridors can be created by topography, e.g. a pass along a mountain ridge, crossed by migratory birds, or by hydrological cycles, e.g. a riverbed, or by human tropical or boreal forest clearing.

Rivers are the most important and extensively studied corridors. Recently Planty-Tabacchi *et al.* (1996) found great numbers of alien species moving along a river corridor. This was explained as a direct effect of an intermediate disturbance regime and the physical structure of the riparian corridor. The patchy structure of the riparian landscape, owing to the combination of seasonal flooding, temporary ponds, and generally extreme environmental conditions, from dry to permanently submerged habitat, allows the presence of both indigenous and alien species. Invasibility along a river depends on the various hydrological and geomorphological zones along that river. The apparent effect of dominance of an alien species on native plants is mitigated by the high landscape temporal heterogeneity of the system, owing to seasonal disturbance.

Corridors seem to be vital for the maintenance of large home range mammals such as the cougar (*Felix concolor*). This species travels on average 5.5 miles per night, and therefore needs a lot of corridors to maintain its living standards. Telemetry seems a very promising approach to understand the mechanism of landscape corridor selection by highly vagile animals: putting this information on a map and processing the data in a GIS is also very promising, although administrative and political restrictions in planning space for cougars and other large carnivores are the real limiting factor to conservation action (Beier 1993).

Using satellite telemetry techniques Morreale *et al.* (1996) recently tracked the oceanic movements of eight female leatherback turtles (*Dermochelis coriacea*) after egg-laying on the beaches of Costa Rica (Fig. 4.24). Apparently all individuals maintained a narrow route between the breeding beach and the open ocean. The existence of a marine corridor is suggested and discussed. In fact, it is not coincidence that different examples in different years were following the same oceanic routes. A gap in these results is caused by the uncertainty of factors determining the corridors, which are probably created by food distribution patterns and the rarity of this turtle could be a signal of decreasing oceanic productivity.

The width of a corridor is an important attribute of these structures (Fig. 4.25), and there is some evidence from an experiment conducted on *Microtus aeconomus* (Pallas) by Andreassen *et al.* (1996) that this species has greater dispersal capacity when intermediate corridors of 0.4 m width are offered (Fig. 4.26).

4.4.3 Summary

- Connectedness and connectivity are two attributes of heterogeneous landscapes.
- Connectedness is the degree of physical connection between patches.
- Connectivity is the process by which subpopulations are interconnected in demographic functional units.
- Corridors are physical or functional narrow belts that increase connectivity and allow the movements of plants and animals in a hostile matrix.
- Although strongly questioned, corridors are important components of the landscape.
- The functions and attributes of corridors are species specific.

4.5 SOIL LANDSCAPE AND MOVEMENT OF WATER AND NUTRIENTS ACROSS LANDSCAPES

4.5.1 Introduction

The surface of the earth shows variability at landscape scale (1–10 km) and this variability has a strong influence on the large-scale circulation of the atmosphere (Pielke and Avissar 1990, Klaassen and Claussen 1995). For instance, the contrast between bare soil and a forest soil can create strong breezes. Surface temperature and precipitation are strongly conditioned by the vegetation cover. Bare soil receives four time less precipitation in North America and the ground surface is 15–25° warmer than in wet soils. Landscape characteristics affect the regional atmosphere and have a strong influence on the global climate. The landscape effect mostly depends on vegetation height and distribution. Heat and gas exchange is sensitive to the roughness of the landscape, and the exchange between the landscape and the atmosphere is not simply the sum of the exchanges between the different landscape elements.

The dispersion of organisms in a landscape has been assumed to be random and unpredictable, assuming that distance from the source is the main factor. In reality the roughness of the landscape, owing to its topographical character and vegetation cover, creates patterns that can be modelled and then predicted. For instance, the circulation patterns of VA (vesicular–arbuscular) mycorrhizal fungi affect the distribution and functioning of plant com-

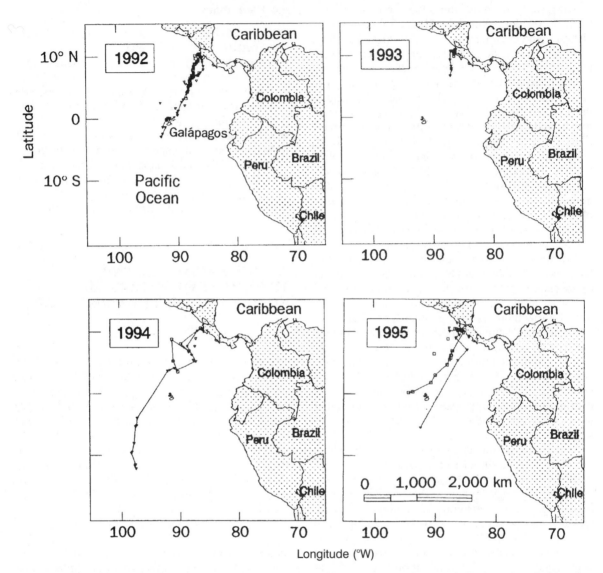

Figure 4.24 Migration routes of leatherback turtles monitored by satellite radio tracking from the Costa Rican breeding beaches to the deep ocean after egg deposition (from Morreale *et al.* 1996, with permission).

munities, altering the succession rate and competition of plants (Allen *et al.* 1989). The deposition of propagules by wind is not related to distance: in fact, it is not decreased as the distance increases, but is more linked to complex dynamics that can be understood if the direction and intensity of the wind are known. We have to understand the physical character of a site, and also the scale at which an

organism reacts with the environment in which selection pressure operates (Fig. 4.27).

4.5.2 Soil landscape

Soil landscapes play a fundamental role in vegetation patterning. This has been clearly demonstrated by McAuliffe (1994), studying landscape evolution,

Figure 4.25 Sibillini Mountains, central Italy. A remnant of beech forest creates a belt along the mountain ensuring a potential corridor for the animals. The recent expansion of the central Italian wolf population (*Canis lupus*) toward the north was probably facilitated by such types of corridors.

soil formation and vegetation distribution in the Sonoran desert bajadas (Arizona).

The Sonoran desert, close to Tucson, is a complex mosaic of distinct geological landforms created by aggradation and erosion of soil in different eras. In this area the soil mosaic has a strong gradient for age and profile development. Vegetation patterns and geomorphic processes are strongly correlated. A fundamental role in controlling vegetation patterns has been found for the weathering intrusive versus weathering-resistant extrusive rocks. Landform age and stability have a strong effect on vegetation. For instance, *Larrea tridentata* occupies many parts of this landscape thanks to a clone-like growth that excludes other species. However, in young alluvial deposits, in highly erodible hillslopes and in extremely thin soil that experiences severe drought due to a petrocalcic horizon (caliche) *L. tridentata*

suffers episodes of mortality. In the more drought-exposed parts of this landscape succulent plant communities can be found. In conclusion, in this landscape soil formation, plant physiological, demographic and interspecies interactions all contribute to create a complex mosaic and permit enlargment of the spatial scale, a relevant framework for studying arid systems. This example can be exported to other regions.

The soil landscape system as conceptualized by Huggett (1975, 1995) is characterized by a dispersive process of all weathering debris (particles, colloids and solutes), influenced by surface and phreatic surface forms. The movement of elements tends to be perpendicular to land-surface forms, altering the topography that in turn influences their movement, creating a feedback between the two systems (mobile debris and topography) (Fig. 4.28).

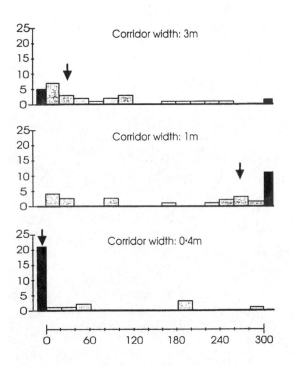

Figure 4.26 Frequency distribution of the maximum distance reached from the release point by males of *Microtus aeconomus* (Palls). The black bars indicates the frequency of males that have never left the release patch or have reached the opposite patch (from Adreassen *et al.* 1996, with permission).

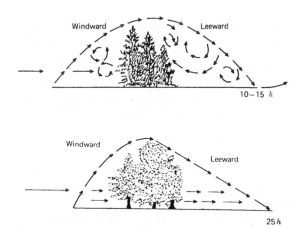

Figure 4.27 Effect of windbreak on wind turbulence. The height, porosity and distance of a windbreak influence the wind's behaviour (from van Eimern *et al.* 1964, with permission).

It is reasonable to expect a strong influence of landscape patterns on soil formation. The role of topography can be appreciated using many descriptors, but the most important seem to be elevation, slope, gradient, slope curvature and length, slope direction, contour curvature and catchment area.

Study of toposequences also indicates that minor variations in topography produce changes in soil properties. Moore *et al.* (1993), studying a toposequence in a Colorado agroecosystem, found a good correlation between slope and wetness index with soil attributes (organic matter content, pH, extractable P, and silt and sand contents), in A horizon thickness, accounting for about 50 per cent of the variability.

Precipitations and their interactions with the landscape depend not only on slope character and soil composition but also on soil cover and use. Figure 4.29 represents the distribution of precipitation according to the different land cover, describing the rate of surface runoff, seepage, evaporation and dissolved matter flow (Ripl 1995).

The landscape position plays a fundamental role, for instance, in the sediment chemistry of abandoned-channel wetlands (Schwarz *et al.* 1996). The riparian landscape is a strongly dynamic system and soil properties are expected to be influenced by this dynamism. The chemistry of riparian wetlands is influenced by the rate of connectivity with the active river channel. In fact, organic matter and nitrogen are in lesser quantity in sites more connected and sensitive to exportation during flooding events. But the neighbouring agricultural mosaic also has a strong influence on the soil chemistry of riparian deposits, especially for nitrogen and phosphorus. The time after abandonment is an important control factor. After abandonment the system moves from being open and dynamic to a closed system in which the circulation of water and materials creates different conditions. The organic matter increases and soil develops, but the nitrogen content seems not to increase significantly with time and the content found in old abandoned deposits is not significantly greater than that found in channels of intermediate age.

Nutrients such as phosphorus and nitrogen are captured and processed differently according to different patch type in a mosaic landscape (Risser 1989). This depends largely on the topographical position of vegetation patches and on edaphic conditions. Nutrients move from one landscape unit to another according to the position of each unit.

Understanding of this process is essential to model and manage ecosystems. Phosphorus movement is strongly linked to particle transportation: any soil particle accumulation process is an indicator of phosphorus trapping capacity. Soil quality plays a fundamental role in nutrient retention and dynamics.

Depending on topography, nutrients such as carbon (C), nitrogen (N) and phosphorus (P) show different concentrations along a soil catena. Schimel et al. (1985) found an increase in C, N and P downslope in a shortgrass steppe (Colorado). The soil properties are different moving from the top downwards in rounded hills of the studied area (Fig. 4.30 and Table 4.3). At the summit the concentration of these three nutrients shows a decrease moving from the top downslope. Backslope is the hilly portion in which C, N and P concentrations are at a minimum. N availability increases downslope, but relative N mineralization decreases.

Another relevant factor affecting the cycle of nutrients is linked to dominant land use of a watershed. The nutrient discharge changes according to the different land use. Table 4.4 reports the quantities in kg/ha of different watersheds (Correll et al. 1992). Nitrogen and phosphorus are mostly discharged by croplands.

4.5.3 The role of riparian vegetation in nutrient dynamics

Peterjohn and Correll (1984) and Correll et al. (1992) have studied the effect of coastal land use and terrestrial community mosaics on nutrient transport to coastal waters. The riparian deciduous hardwood forest bordering fields removes over 80% of nitrate and total phosphorus in overland flooding, and about 85% of nitrate in shallow groundwater drainage from cropland. But the nutrient discharge from croplands is higher than the discharge from pastures and other forests. Estuarine tidal marshes capture organic material and release dissolved nutrients (Fig. 4.31).

Table 4.4 Annual discharge of nutrients in three different watersheds dominated by different land cover. The nutrients are in kg/ha (From Correll et al. 1992, with permission)

Parameter	Cropland	Pasture	Forest
Total nitrogen	13.80	5.95	2.74
Dissolved ammonium	0.45	0.51	0.15
Nitrate	6.35	3.20	0.36
Total phosphorus	4.16	0.68	0.63
Orthophosphate	1.20	0.32	0.15
Atomic ratio of total nitrogen:total phosphorus	1.50	3.95	1.96

Different zones of bed structure and stability in a channel control the pattern of bed load transport at fine scale, but in the medium and long term the medium bed load yield has a low variability. Local bed load transport is influenced by local conditions in the channel. The spatial pattern of flow competence and bed load transport is more complex than would be expected from a simple relationship with shear stress (Powell and Ashworth 1995).

The riparian forest has an important role in regulating the upstream–downstream movement of matter and energy. Because geomorphic processes determine the structure of channels and floodplains, they also influence the capacity of soil to retain nutrients and organic carbon. Two erosional types (E-type characterized by sand deposition and D-type characterized by silt and clay deposition) retain nutrients in different ways. C, N and P are correlated with erosion/sedimentation processes. D-type riparian forest soil behaves as a sink during flood periods; E-type riparian forest functions as a source, releasing large amounts of C, N and P (Pinay et al. 1992). This demonstrates that a riparian forest functions not as a homogeneous buffer, but depending on different geomorphic processes the same forest type can have different nutrient-

Table 4.3 Organic C, N and P concentrations in catena soil surface horizons (from Schimel et al. 1985, with permission)

Slope orientation	Horizon	Concentration (mg/kg)		
		$C_{organic}$	$N_{organic}$	$P_{organic}$
Summit	A	6900±800	921±97	124±25
Backslope	A	5700±400	665±92	65±16
Footslope	Ai	209400±1500	1937±275	206±61

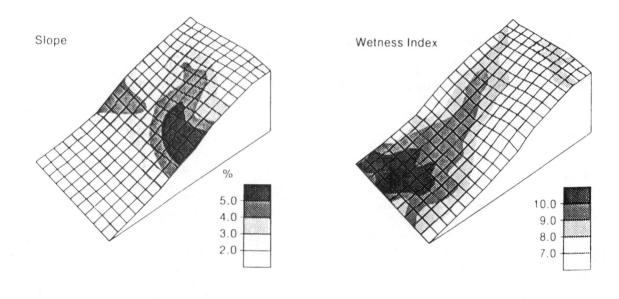

Figure 4.28 Slope, wetness index, A horizon, P, organic matter and pH measured and predicted using a 15.24 m grid-based digital elevation model of the Sterling, CO, site (from Moore *et al.* 1993, with permission).

retention capacities according to the heterogeneity of the patch mosaic.

Kesner and Meentemeyer (1989), studying the Little River Watershed, Georgia, USA, found a massive flow of N, especially from anthropogenic sources, but the total balance of N indicates that a buffering process occurs, mainly due to riparian vegetation, despite a large agricultural input. Once again the importance of riparian forest is confirmed, with strong implications for cropland management.

4.5.4 Origin, composition and flux of dissolved organic carbon in a small watershed

Dissolved organic carbon (DOC) has been studied in different locations in the Hubbard Brook Valley

by McDowell and Likens (1988). Of special interest from a landscape point of view seems to be the question posed by these authors: Are there significant changes in the composition of DOC as it moves through the landscape?

The water flowing across the Hubbard Brook landscape shows an increase in DOC with passage through the upper soil horizons. By a leaching process water is enriched by DOC moving first through the canopy then through the forest floor. DOC remains constant moving from the forest floor to streams and then into the Mirror Lake. DOC decreases on passing through the mineral soil, but carbohydrates increase in the sites of high primary production, especially in throughfall and lake water where photosynthesis is very high (Fig. 4.32).

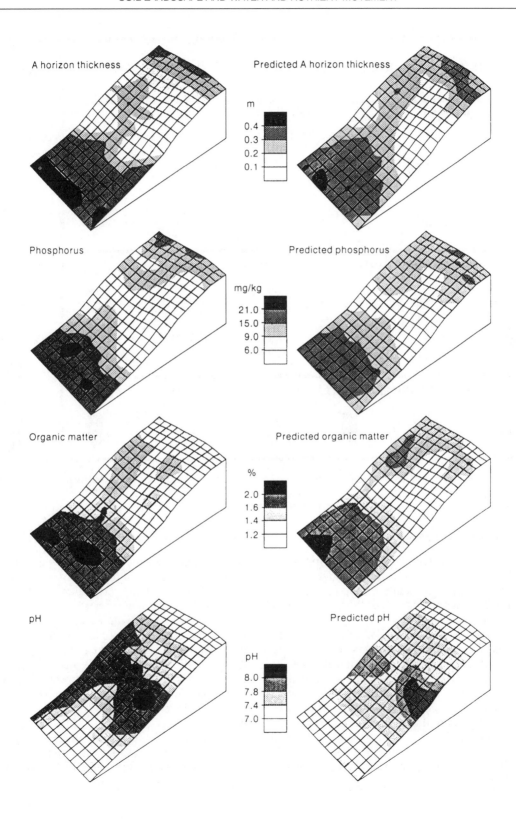

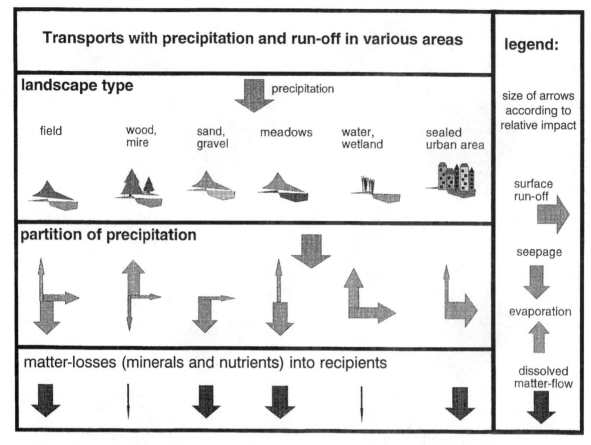

Figure 4.29 Precipitation and matter losses across different components of a landscape (from Ripl 1995, with permission).

4.5.5 Leaf litter movement in the landscape

In deciduous forests the leaf litter is an important source of nutrients, and their movement across different patches depends mainly on topography. The redistribution of leaves influences the heterogeneity of a landscape.

The behaviour of litter fall depends greatly on the orientation of slopes and on tree species. Boerner and Kooser (1989) found that net downslope litter movement was larger than vertical litter fall. *Quercus* litter was 1.3–1.5 times more redistributed than non-*Quercus* litter, and most of the redistribution occurred during the January–April leafless season. A different behaviour was observed for different slopes and orientations. This in general contributes to maintaining fertility patchiness in the landscape.

4.5.6 Spatial patterns of soil nutrients

Nutrients are present in the soil with a heterogeneous distribution, and this pattern is especially evident in the pre- and desert regions where nutrients are scarce.

In the last century most of the deserts of the southwest United States have changed their main land cover from grasses to shrubs. The causes of this dramatic change, largely due to overgrazing, appear different and not very clear, but once a shrub has grown in one location within a short time an 'island of fertility' is created (Schlesinger *et al.* 1995).

Nutrients like N in a perennial grassland (*Beteloua eriopoda*) show variations of 35–76% at distances <20 cm; the remaining variance was expressed over a distance of 7 m.

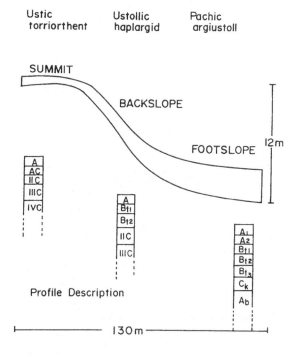

Figure 4.30 Different soil horizon along a toposequence in a short-grass steppe catena (from Schimel *et al.* 1985, with permission).

In adjacent shrubland in which *Larrea tridentata* has replaced grasses, N variation stretches from 1.0 to 3.0 m and seems more concentrated under shrub canopy. This accumulation has been found also for soil PO_4, Cl, SO_4 and K. In the intershrub spaces are greater concentrations of Rb, Na, Li, Ca, Mg and Sr.

In conclusion, grasslands show a more fine-grained distribution of soil properties but shrublands have a more coarse-grained distribution of these components. The soil properties could be utilized in semidesert regions as an index of desertification, from grassland across shrublands.

4.5.7 Summary

- Topography and atmospheric circulation are linked by complex feedbacks at different scales.
- Wind dispersion capacity is modified by landscape topography, affecting in turn the distribution and dispersion of anemochore organisms such as fungi.
- Vegetation patterns and geomorphic processes are strictly correlated.

- Minor variations in topography produce changes in soil properties, which in turn affect vegetation type and distribution.
- Soil nutrients have different behaviours according to the position in the landscape and the land cover.
- Riparian vegetation has an important filtering role for C, N and P moving along the slope discharge.
- Dissolved organic carbon (DOC) increases moving through a landscape owing to the leaching process across forest canopies and the upper soil profile, but seems constant on the forest floor.
- Landscape topography is important in determining the fate of leaf litter movement, which is a fundamental source of nutrients in forests.
- Fine- and coarse-grained distribution of soil chemistry may be a good indicator of desertification in semiarid regions.

SUGGESTED READING

Butler, D.R. *Zoogeomorphology. Animals as geomorphic agents.* Cambridge University Press, Cambridge, 1995.

Goldammer, J.G. and Jenkins, M.J. (eds.) *Fire in ecosystem dynamics. Mediterranean and northern perspectives.* SPB Academic Publishing, The Hague, 1990.

Harris, L.D. *The fragmented forest. Island biogeography theory and the preservation of biotic diversity.* University of Chicago Press, Chicago, 1984.

Huggett, R.J. *Geoecology. An evolutionary approach.* Routledge, London, 1985.

Kozlowski, T.T. and Ahlgren, C.E. (eds.) *Fire and ecosystems.* Academic Press, New York, 1974.

MacArthur, R.H. and Wilson, E.O. *The theory of island biogeography.* Princeton University Press, Princeton, 1967.

Miller, D.E. *Energy at the surface of the earth. An introduction to the energetics of ecosystems.* Academic Press, New York, 1981.

O'Neill P. *Environmental chemistry.* Chapman & Hall, London, 1985.

Pickett, T.A. and White, P.(eds.) *The ecology of natural disturbance and patch dynamics.* Academic Press, Orlando, 1985.

Trabaud, L. (ed.) *The role of fire in ecological systems.* SPB Academic Publishing, The Hague, 1987.

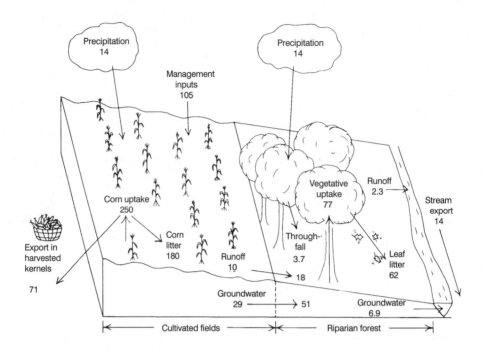

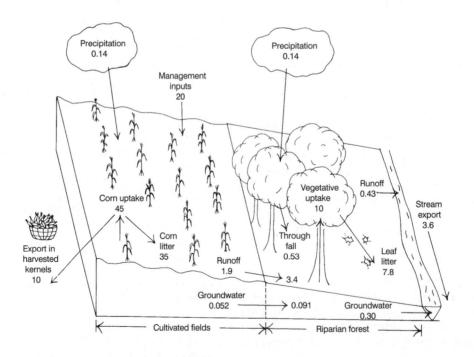

Figure 4.31 Diagram of total N and total P flux and cycle in the period March 1981 to March 1982 in a small watershed, Rhode River drainage basin, Maryland (from Peterjohn and Correll 1984, with permission).

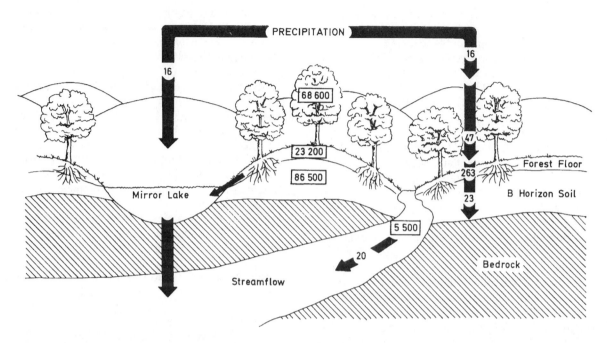

Figure 4.32 Movements and standing stock of organic carbon and flux of dissolved organic carbon (DOC) in the Bear Brook watershed (from MacDowell and Likens 1988, with permission).

REFERENCES

Aizen, M.A. and Feinsinger, P. (1994) Habitat fragmentation, native insect pollinators, and feral honey bees in Argentine Chaco Serrano. *Ecological Applications* **4**: 378–392.

Allen, M.F., Hipps, L.E., Wooldridge, G.L. (1989). Wind dispersal and subsequent establishment of VA mycorrhizal fungi across a successional arid landscape. *Landscape Ecology* **2**: 165–171.

Andreassen, H.P., Halle, S., Ims, R.A. (1996) Optimal width of movement corridors for root voles: not too narrow and not too wide. *Journal of Applied Ecology* **33**: 63–70.

Andren, H., Angelstam, P., Lindtstrom, E., Widen, P. (1985) Differences in predation pressure in relation to habitat fragmentation: an experiment. *Oikos* **45**: 273–277.

Baker, W.L. (1992) The landscape ecology of large disturbances in the design and management of nature reserves. *Landscape Ecology* **7**: 181–194.

Bancroft, G.T., Strong, A.M., Carrington, M. (1995) Deforestation and its effects on forest-nesting birds in the Florida Keys. *Conservation Biology* **9**: 835–844.

Baudry, J. (1984) Effects of landscape structures on biological communities: the case of hedgerow network landscapes. In: Brandt, J. and Agger, P. (eds.). *Methodologies in landscape ecological research and planning.* Vol. 1 Proceedings of the First International Seminar of the International Association of Landscape Ecology, Roskilde, Denmark, October 15–19, pp. 55–65.

Baudry, J. and Merriam, G. (1987) Connectivity and connectedness: functional versus structural patterns in landscapes. In: Schreiber, K.F. (ed.) *Connectivity in landscape ecology.* Proceedings of the 2nd International Seminar of the International Association for Landscape Ecology. *Munstersche Geographische Arbeiten* **29**, 1988, pp. 23–28.

Beier, P. (1993) Determining minimum habitat areas and habitat corridors for cougars. *Conservation Biology* **7**: 94–108.

Belsky, A.J. and Canham, C.D. (1994) Forest gaps and isolated savanna trees. *Bioscience* **44**: 77–84.

Bierregaard, R.O. Jr., Lovejoy, T.E., Kapos, V., dos Santos, A.A., Hutchings, R.W. (1992) The biological dynamics of tropical rainforest fragments. *BioScience* **42**: 859–866.

Blake, J.G. (1986) Species–area relationship of migrants in isolated woodlots in east-central Illinois. *Wilson Bulletin* **98**: 291–296.

Blake, J.G. and Hoppes, G. (1986) Influence of resource abundance on use of treefall gaps by birds. *Auk* **103**: 328–340.

Blake, J.G. and Karr, J.R. (1987) Breeding birds of isolated woodlot: area and habitat relationships. *Ecology* **68**: 1724–1734.

Bleich, V.C., Wehausen, J.D., Holl, S.A. (1990) Desert-dwelling mountain sheep: conservation implications of a naturally fragmented distribution. *Conservation Biology* **4**: 383–390.

Boerner, R.E. and Kooser, J.G. (1989) Leaf litter redistribution among forest patches within an Allegheny Plateau watershed. *Landscape Ecology* **2**: 81–92.

Brothers, T.S. and Spingarn, A. (1992) Forest fragmentation and alien plant invasion of central Indiana old-growth forests. *Conservation Biology* **6**: 91–100

Burger, L.D., Burger, L.W., Faaborg, J. (1994) Effects of prairies fragmentation on predation on artificial nests. *Journal of Wildlife Management* **58**: 249–254.

Burkey, T.V. (1995) Extinction rates in archipelagoes: implications for population in fragmented habitats. *Conservation Biology* **9**: 527–541.

Castello, J.D., Leopold, D.J., Smallidge, P.J. (1995) Pathogens, patterns, and processes in forest ecosystems. *Bioscience* **45**: 16–24.

Correll, D.L., Jordan, T.E., Weller, D.E. (1992) Nutrient flux in a landscape: effects of coastal land use and terrestrial community mosaic on nutrient transport to coastal waters. *Estuaries* **15**: 431–442.

Cutler, A. (1991) Nested faunas and extinction in fragmented habitats. *Conservation Biology* **5**: 496–505.

Diamond, J.M. (1969) Avifaunal equilibria and species turnover rates on the Channel Islands of California. *Proceedings of the National Academy of Sciences, USA* **69**: 3199–3203.

Diffendorfer, J.E., Gaines, M.S., Holt, R.D. (1995) Habitat fragmentation and movements of three small mammals (*Sigmodon, Microtus,* and *Peramyscus*). *Ecology* **76**: 827–839.

Gibbs, J.P. and Faarborg, J. (1990) Estimating the viability of ovenbird and kentucky warbler populations in forest fragments. *Conservation Biology* **4**: 193–196.

Gluck, M.J. and Rempel, R.S. (1996) Structural characteristics of post-wildfire and clearcut landscapes. *Environmental Monitoring and Assessment* **39**: 435–450.

Herkert, J.R. (1994) The effect of habitat fragmentation on midwestern grassland bird communities. *Ecological Applications* **4**: 461–471.

Hinsley, S.A., Bellamy, P.E., Newton, I. (1995) Bird species turnover and stochastic extinction in woodland fragments. *Ecography* **18**: 41–50.

Hobbs, R.J. and Hopkins, A.J.M. (1990) From frontier to fragments: European impact on Australia's vegetation. *Proceedings of the Ecological Society of Australia* **16**: 93–114.

Hobbs, R.J. and Mooney, H.A. (1991) Effects of rainfall variability and gopher disturbance on serpentine annual grassland dynamics. *Ecology* **72**: 59–68.

Huggett, R.J. (1975) Soil landscape systems: a model of soil genesis. *Geoderma* **13**: 1–22.

Huggett, R.J. (1995) *Geoecology. An evolutionary approach.* Routledge, London and New York.

Kattan, G.H., Alvarez-Lopez, H., Giraldo, M. (1994) Forest fragmentation and bird extinctions: San Antonio eighty years later. *Conservation Biology* **8**: 138–146.

Kesner, B.T. and Meentemeyer, V. (1989) A regional analysis of total nitrogen in an agricultural landscape. *Landscape Ecology* **2**: 151–163.

Klaassen, W. and Claussen, M (1995) Landscape variability and surface flux parameterization in climate models. *Agricultural and Forest Meteorology* **73**: 181–188.

Klein, B.C. (1989) Effects of forest fragmentation on dung and carrion beetle communities in central Amazonia. *Ecology* **70**: 1715–1725.

Knick, S.T. and Rotenberry, J.T. (1995) Landscape characteristics of fragmented shrubsteppe habitats and breeding passerine birds. *Conservation Biology* **9**: 1059–1071.

Leimgruber, P., McShea, W.J., Rappole, J.H. (1994) Predation on artificial nests in large forest blocks. *Journal of Wildlife Management* **58**: 254–260.

Lertzman, K.P., Sutherland, G.D., Inselberg, A., Saunders, S.C. (1996) Canopy gaps and the landscape mosaic in a coastal temperate rain forest. *Ecology* **77**: 1254–1270.

Lord, J.M. and Norton, D.A. (1990) Scale and the spatial concept of fragmentation. *Conservation Biology* **4**: 197–202.

MacArthur, R.H. and Wilson, E.O. (1967) *The theory of island biogeography*. Princeton University Press, Princeton, NJ.

McAuliffe, J. R. (1994) Landscape evolution, soil formation, and ecological patterns and processes in Sonoran desert Bajadas. *Ecological Monographs* **64**: 111–148.

McDowell, W.H. and Likens, G.E. (1988) Origin, composition, and flux of dissolved organic carbon in the Hubbard Brook valley. *Ecological Monographs* **58**: 177–195.

Margules, C.R., Milkovits, G.A., Smith, G.T. (1994) Contrasting effects of habitat fragmentation on the scorpion *Cercophonius squama* and an amphipod. *Ecology* **75**: 2033–2042.

Martin, J.L. , Gaston, A.J., Hitier, S. (1995) The effect of island size and isolation on old growth forest habitat and bird diversity in Gwaii Haanas (Queen Charlotte Islands). *Oikos* **72**: 115–131.

Merriam, G. (1984) Connectivity: a fundamental ecological characteristic of landscape pattern. In: Brandt, J. and Agger, P. (eds.). *Methodologies in landscape ecological research and planning*. Vol. 1 Proceedings of the first international seminar of the International Association of Landscape Ecology. Roskilde, Denmark, October 15–19. Pp. 5–15.

Moloney, K.A. and Levin, S.A. (1996) The effect of disturbance architecture on landscape-level population dynamics. *Ecology* **77**: 375–394.

Moore, I.D., Gessler, P.E., Nielsen, G.A., Peterson, G.A. (1993) Soil attribute prediction using terrain analysis. *Soil Science Society of America Journal* **57**: 443–452.

Moore, P.D. (1996) Fire damage soils our forests. *Nature* **384**: 312–313.

Morreale, S.J., Standora, E.A., Spotila, J.R., Paladino, F.V. (1996) Migration corridor for sea turtles. *Nature* **384**: 319–320.

Naveh, Z. (1990) Fire in the Mediterranean – a landscape ecological perspective. In: Goldhammer, J.G. and Jenkins, M.J. (eds.), *Proceedings of the Third International Symposium on Fire Ecology*, Freiburg, FRG, 1989. SPB Academic Publishing, The Hague, pp. 1–20.

Naveh, Z. (1991) The role of fire in Mediterranean vegetation. *Botanic Chronicle* **10**: 385–405.

Naveh, Z. (1992) A landscape ecological approach to urban systems as part of the total human ecosystems. *Journal of the Natural History Museum Institute, Chiba* **2**: 47–52.

Newmark, W.D. (1990) Tropical fragmentation and the local extinction of understory birds in the eastern Usambara Mountains, Tanzania. *Conservation Biology* **5**: 67–78

Nilsson, C. and Grelsson, G. (1995) The fragility of ecosystems: a review. *Journal of Applied Ecology* **32**: 677–692.

Pasitschniak, M. and Messier, F. (1995) Risk of predation on waterfowl nests in the Canadian prairies: effects of habitat edges and agricultural practices. *Oikos* **73**: 347–355.

Peterjohn, W.T. and Correll, D.L. (1984) Nutrient dynamics in an agricultural watershed: observations on the role of a riparian forest. *Ecology* **65**: 1466–1475.

Pickett, S.T.A. and White, P.S. (1985) Patch dynamics: a synthesis. In: Pickett, T.A. and White, P. (eds.), *The Ecology of natural disturbance and patch dynamics*. Academic Press, Orlando, pp. 371–384.

Pielke, R.A. and Avissar, R. (1990) Influence of landscape structure on local and regional climate. *Landscape Ecology* **4**: 133–155.

Pinay, G., Fabre, A., Vervier, Ph., Gazelle, F. (1992) Control of C, N, P distribution in soils of riparian forests. *Landscape Ecology* **6**: 121–132.

Planty-Tabacchi, A-M, Tabacchi, R., Naiman, R.J., Deferrari, C., Décamps, H. (1996) Invasibility of species-rich community in riparian zones. *Conservation Biology* **10**: 598–607.

Poorter, L., Jans, L., Bongers, F., Van Rompaey, R.S.A.R. (1994) Spatial distribution of gaps along three catenas in the moist forest of Tai Park, Ivory Coast. *Journal of Tropical Ecology* **10**: 385–398.

Powell, D.M. and Ashworth, P.J. (1995) Spatial pattern of flow competence and bed load transport in a divided gravel bed river. *Water Resource Research* **31**: 741–752.

Redpath, S.M. (1995) Impact of habitat fragmentation on activity and hunting behaviour in the tawny owl, *Strix aluco*. *Behavioural Ecology* **6**: 410–415.

Ripl, W. (1995) Management of water cycle and energy flow for ecosystem control: the energy–transport–reaction (ETR) model. *Ecological Modelling* **78**: 61–76.

Risser, P.G. (1989) The movement of nutrients across heterogeneous landscapes. In: M. Clarholm and L. Bergstrom (eds.) *Ecology of arable land*. Kluwer Academic Publishers, Dordrecht, pp. 247–251.

Roland, J. (1993) Large-scale forest fragmentation increases the duration of tent caterpillar outbreak. *Oecologia* **93**: 25–30.

Rudis, V.A. (1995) Regional forest fragmentation effects on bottomland hardwood community types and resource values. *Landscape Ecology* **10**: 291–307.

Saunders, D.A., Hobbs, R.J., Margules, C.R. (1991) Biological consequences of ecosystem fragmentation: a review. *Conservation Biology* **5**: 18–32.

Schieck, J., Lertzman, K., Nyberg, B., Page, R. (1994) Effects of patch size on birds in old-growth montane forests. *Conservation Biology* **9**: 1072–1084.

Schimel, D., Stillwell, M.A., Woodmansee, R.G. (1985) Biogeochemistry of C, N, P in a soil catena in the short-grass steppe. *Ecology* **66**: 276–282.

Schlesinger, W.H. , Raikes, J.A., Hartley, A.E., Cross, A.F. (1995) On the spatial pattern of soil nutrients in desert ecosystems. *Ecology* **77**: 364–374.

Schwarz, W.L., Malanson, G.P., Weirich, F.H. (1996) Effect of landscape position on the sediment chemistry of abandoned-channel wetlands. *Landscape Ecology* **11**: 27–38.

Simberloff, D., Farr, J.A., Cox, J., Mehlman, D.W. (1992) Movement corridors: conservation bargains or poor investment? *Conservation Biology* **6**: 493–504.

Steinauer, E.M. and Collins, S.L. (1995) Effects of urine deposition on small-scale patch structure in prairie vegetation. *Ecology* **76**: 1195–1205.

Stouffer, P.C. and Bierregaard, R.O. (1995) Use of amazonian forest fragments by understory insectivorous birds. *Ecology* **76**: 2429–2445.

Tscharntke, T. (1992) Fragmentation of *Phragmites* habitats, minimum viable population size, habitat suitability, and local extinction of moths, midges, flies, aphids, and birds. *Conservation Biology* **6**: 530–536.

van Eimern, J., Karschon, R. , Razumova, L.A., Robertson, G.W. (1964) *Wind-breaks and shelterbelts*. WMO Tech. Note 59, 188 pp.

Villard, M.-A. and Taylor, P.D. (1994) Tolerance to habitat fragmentation influences the colonization of new habitat by forest birds. *Oecologia* **98**: 393–401.

Villard, M.-A., Merriam, G., Maurer, B.A. (1995) Dynamics in subdivided populations of neotropical migratory birds in a fragmented temperate forest. *Ecology* **76**: 27–40.

Walker, D.A., Halfpenny, J.C., Walker, M.D., Wessman, C.A. (1993) Long-term studies of snow–vegetation interactions. *Bioscience* **43**: 287–301.

White, P.S. and Pickett, T.A. (1985) Natural disturbance and patch dynamics: an introduction. In: Pickett, T.A. and White, P. (eds.), *The ecology of natural disturbance and patch dynamics*. Academic Press, Orlando.

Wiens, J.A. (1994) Habitat fragmentation: island v landscape perspectives on bird conservation. *Ibis* **137**: S97–S104.

Wilcove, D.S. (1985) Nest predation in forest tracts and the decline of migratory songbirds. *Ecology* **66**: 1211–1214.

Wilcove, D.S., McLellan, C.H., Dobson, A.P. (1986) Habitat fragmentation in the temperate. In: Soulé, M.E. (ed.), *Conservation biology*. Sinauer Associates Inc., pp. 237–256.

Willson, M.F., De Santo, T., Sabag, C., Armesto, J.J. (1994) Avian communities of fragmented south temperate reainforests in Chile. *Conservation Biology* **8**: 508–520.

Wunderle, J.M. Jr (1995) Responses of bird populations in a Puerto Rican forest to hurricane Hugo: the first 18 months. *The Condor* **97**: 879–896.

Emerging patterns in the landscape

5.1 INTRODUCTION

Landscapes are complicated systems that show many different patterns according to the scale of resolution and the components towards which the investigation is directed.

Two main patterns are described in this chapter, heterogeneity and ecotones. Both are produced by different processes, of which disturbance and fragmentation in particular are very influential.

Heterogeneity is the main pattern in every landscape, and strictly linked with this pattern we find the edges or ecotones. Every pattern produced by a process is in turn a producer of new processes. In particular, heterogeneity has a dramatic influence on many processes across the landscape.

Ecotones are special areas in which different types of habitats meet and in which ecological processes are strongly influenced by the co-occurrences of different land attributes. Their role is fundamental to an understanding of landscape complexity and function.

5.2 LANDSCAPE HETEROGENEITY

5.2.1 Introduction

Most ecosystem-oriented studies have introduced a basal bias extrapolating the ecosystem from the real world, considering for convenience only homogeneous and quite simple systems. Recently environmental heterogeneity has captured the attention of many scientists, with an incredible pro-duction of new data at any scaled level from individual to landscape (e.g. Pickett and White 1985, Turner 1987, Hastings 1990, Shorrock and Swingland 1990, Kolasa and Pickett 1991). Most actual landscape ecology is geared to the study of phenomena interconnected with or conditioned by spatial heterogeneity.

Heterogeneity is an inherent character of the land mosaic; it exists at any scale of resolution and can be considered as the structural substrate on which biological diversity can develop. Heterogeneity may be defined as the uneven, non-random distribution of objects (Forman 1995), and the analysis of this pattern is of fundamental importance to understanding most ecological processes and the functioning of complicated systems such as landscapes.

Heterogeneity and diversity are two related concepts in landscape ecology, but whereas diversity describes the different qualities of the patches, heterogeneity represents the spatial complexity of the mosaic.

At least three different types of heterogeneity have to be considered:

1. Spatial heterogeneity. This may be seen as a static or a dynamic pattern (Kolasa and Rollo 1991). Spatial heterogeneity has effects on many ecological processes, such as soil formation, weathering, plant distribution, animal distribution, abundance and movements, water and nutrient fluxes, energy storing and recycling etc. Spatial heterogeneity may be divided into horizontal and vertical components. **Horizontal heterogeneity** represents the uneven distribution of

land cover that may be created by human disturbance regimes. For instance, in the Mediterranean basin most of the landscapes have been shaped by a plurimillenary human disturbance regime by which the heterogeneity is enhanced by the diverse cultivations. **Vertical heterogeneity** represents the uneven distribution of vegetation above ground, and is more connected with natural landscapes.

2. Temporal heterogeneity has a meaning similar to spatial heterogeneity, but is measured as a variation at one point in space for different times. Two locations may have identical temporal patterns but be asynchronous in time, thereby expressing temporal heterogeneity.

3. Functional heterogeneity is the heterogeneity of ecological entities (distribution of individuals, populations, species, communities). It may be linked to the life history of organisms at several scales.

Heterogeneity can be observed in the soil composition, as highlighted by Becher (1995), who showed that pH, C_{org}, C_{carb}, texture, saturated hydraulic conductivity, bulk density and pore size distribution vary strongly across plots. This variation can be found at all spatial scales, from landscape to hundreds or tens of metres. This variability can be found also in the vertical soil profile.

The effect of geological heterogeneity creates a highly unpredictable system, and for this reason a probabilistic approach is used by Hantush and Marino (1995) to evaluate aquifer behaviour.

Heterogeneity creates borders, edges and contrast between different patches, and this pattern creates more new processes, influencing, for example, the movement of organisms and fluxes of material and energy (Pickett and Cadenasso 1995).

Plant and animal assemblages can react in a very short time to any change in mosaic heterogeneity, and this can easily be detected by field and/or remote sensing investigations.

Heterogeneity is also a sign of land patchiness. The level of heterogeneity can negatively affect some processes. For instance, when open spaces such as prairies are too small some species of birds (e.g. skylarks) avoid them, although the character of these open spaces in terms of vegetation and resource availability is apparently very close to sites of larger size. Heterogeneity, in consequence, plays a pre-eminent role, especially when the interacting scale of the organisms is coarse.

Heterogeneity may initiate or exaggerate biological interactions with the environment. Local uniqueness, determined by local character and by past site-related history or distinctiveness, is a relevant contributor to spatial heterogeneity. Dispersal is another relevant factor (Levin 1976).

Heterogeneity can admit multiple stable states as a polyclimax, which can be considered a state of spatial variation in the local equilibrium. Heterogeneity is conditioned not only by external factors, such as variations in weather, climate and edaphic factors, but also by internal random events, such as outbreaks, colonization etc.

At this level disturbance can be a relevant factor, interpreting successional processes that to some extent modify or rejuvenate secondary succession, interrupting successional stages and avoiding the formation of homogeneous steady states. The phase difference or maturity is the time lag since the last disturbance event. Other local random events can overlap phase differences, creating a more diverse environment. Also, a moderate disturbance regime increases heterogeneity but this behaviour is quite different according to the starting status of the environment (Fig. 5.1).

5.2.2 Scale and ecological neighbourhoods

Heterogeneity is a concept that can be applied to any scale of the landscape. A system can have higher or lower heterogeneity, depending on the resolution at which it is observed.

A patch may be defined as a discontinuity in environmental character states. Environmental patterning assumes a great importance for most ecological processes. Heterogeneity means that there are two different patch types, which can be different in suitability. Three main categories of spatial aggregation have been recognized by Addicot *et al.* (1987): divided homogeneous, undivided heterogeneous and divided heterogeneous (Fig. 5.2).

In the first case we can assume that suitable patches for a species are embedded in a matrix of an unsuitable medium, such as the sea for insects. In the second and third cases undivided heterogeneity shows a different quality of patches, and divided heterogeneity has patches of different quality dispersed in an unsuitable environment.

For example, fire, grazing and a combination of the two were tested at local and regional scale in the tallgrass prairies of northeastern Kansas by Glenn *et*

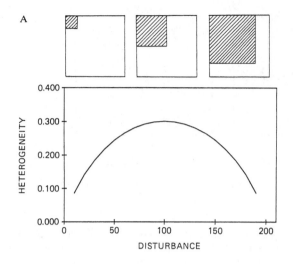

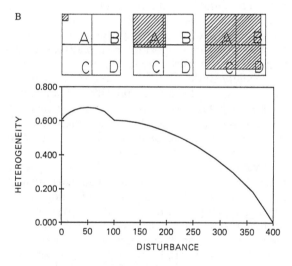

Figure 5.1 Habitat heterogeneity as a function of spatial scale of disturbance. (**A**) Diffusion of disturbance in a homogeneous habitat; (**B**) in a heterogeneous habitat (from Kolasa and Rollo 1991, with permission).

al. (1992). The disturbance had different effects according to the spatial scale: burning produced more heterogeneity at a local scale, and grazing seemed more efficient at a regional scale (15×15 km). At a local scale (0.1 ha) the undisturbed control plot was more heterogeneous than the treated plots. Regional responses to disturbance were more unpredictable than local responses. Spring burning opened dense grass cover, facilitating the germina-

tion of seeds. Grazing after burning maintained open spaces, favouring the implantation of rare species. Burning in the fall reduced species diversity, and this was probably dependent on the poorer seed bank.

The greater heterogeneity at a regional scale of burned or grazed + burned treatments probably depended on the effect of species dynamism in time and space.

The patchy character of an environment creates more difficulties in the choice of the spatial and temporal scale for investigation. Patchiness is not always important for all ecological processes, and the responses of organisms seem an appropriate criterion by which to scale the environment.

In order to find a criterion to size environmental patterning it seems useful to introduce the concept of neighbourhood as the region within which an organism is active or has some influence in an appropriate period of time. The neighbourhood dimension is linked to the organisms and to the ecological process that has been selected for comparison. Using the neighbourhood concept it is possible to measure the relative size, isolation and duration of patches.

Fine or coarse grain is a matter of neighbourhood characteristics. So, if a patch is large for a neighbourhood, then an organism will use that patch in a coarse way. A great variety of indicators can be used to estimate neighbourhood size, but for vagile organisms the net movement of individuals seems appropriate. For sessile organisms the neighbourhood size can be estimated according to the regions from which food, predators and mutualistic foragers come. This approach seems very promising because scaling the neighbourhood means scaling the processes that interact with an organism.

5.2.3 Disturbance and heterogeneity

According to Risser (1987) the role of heterogeneity in the disturbance regime can be controversial. In fact, in some cases heterogeneity can interrupt the spreading of a disturbance, such as a fire in a mixed woodland in which the inflammability of coniferous species is higher than that of broadleaves. In other cases, such as in agroecosystems, in which woodlots are interdispersed in a matrix of cultivation, deer, which find suitable habitats in woodlot, can disturb the surrounding crops.

The non-random distribution of species across scales mainly depends on the community heterogeneity (Collins 1992). Larger samples include a

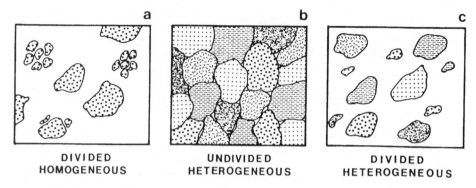

Figure 5.2 Possible combinations of patchiness in a heterogeneous landscape (from Addicot *et al.* 1987, with permission).

higher number of species than small samples, so we expect a greater similarity in large plots than in small ones.

Heterogeneity was negatively correlated with burning frequency in the Konza Prairies (Collins 1992) (Figs 5.3 and 5.4). The mean annual heterogeneity on annual burned grasslands was always less than on unburned plots. In this case an intermediate disturbance regime decreases heterogeneity. Spatial and temporal heterogeneity are positively correlated at one-year and four-year burning frequencies.

5.2.4 Heterogeneity and animals

Levins (1968), in his book *Evolution in changing environments*, dealt mainly with the genetic consequences of a patchy environment on the way in which an animal perceives the environment as fine or coarse grained.

Spatial heterogeneity, measured as horizontal variability in the type profiles of a habitat, has been recognized as one of the major factors influencing bird species diversity (BSD) (MacArthur *et al.* 1962).

Roth (1976) found a good relationship between an index of heterogeneity and BSD, but as the trees became denser the canopy had less influence on heterogeneity. Scattered trees and shrubs were more important in ensuring habitat patchiness. However, percent overlap (PCO), i.e. the percentage of the entire community which is expected to overlap at a given site, is negatively correlated with BSD (Roth 1976) (Figs 5.5 and 5.6).

Patchiness can be used to explain the differences in diversity between sites in the same habitats: the higher bird diversity in shrubby habitats probably depends on the higher shrub heterogeneity, compared with low diversity in tree habitats, although these have more vegetation layers or greater volume. It is also quite clear that vegetation physiognomy is not enough to explain animal distribution, but that floristic composition also plays an important role (Rotenberry 1985).

Recently Butler (1995) reviewed the role of animals as geomorphic agents and, although no explicit reference has been made to heterogeneity, it appears quite clear from his presentation that animal effects have increasingly to be considered as factors of spatial heterogeneity.

Invertebrates such as ants and termites, vertebrates such as shore birds, and mammals such as beaver, buffalo, rodents and moles are building, burrowing, digging , trampling and moving soil and opening up the vegetation cover. Domestic livestock, when densely reared, can permanently modify the vegetation cover, producing a new landscape. An example is the terracettes produced on steep slopes on mountain pasture by sheep and goats trampling.

The savanna vegetation in Zimbabwe is patterned at various scales (Scoones 1995). At the regional scale are rainfall gradient and geomorphology. At the landscape scale is the dispersal of soil types. At the microscale vegetation heterogeneity is dominated by slope, soil type and vegetation disturbance regimes. To maintain resources in savanna grazed by cattle, wild animals and goats, it is important to understand the mechanisms which, working at different scales, have the capacity to reduce competition and maintain the heterogeneity of the vegetation patches, thereby ensuring the resilience of the system. Scoones argues for the

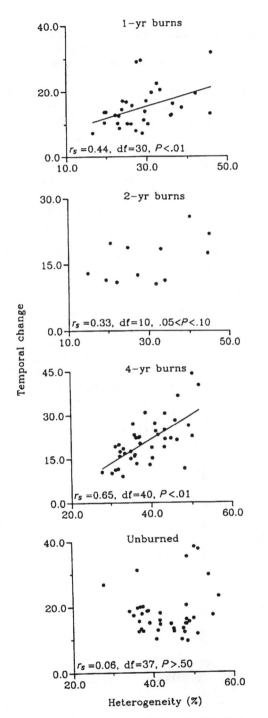

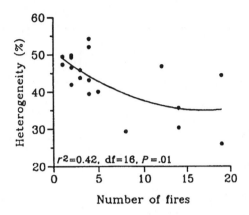

Figure 5.4 Relationship between number of times a site has been burned between 1972 and 1990, and site heterogeneity in 1990 (from Collins 1992, with permission).

necessity to ensure opportunistic and flexible movements at different scales between resources.

5.2.5. Spatial heterogeneity and prey–predator control systems

Spatial heterogeneity can be responsible for some spatial density dependence (Dempster and Pollard

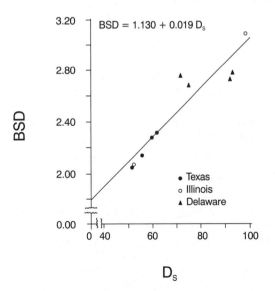

Figure 5.3 Spearman's rank correlation of within-site heterogeneity in species composition, in a given year, and proportional change in species composition (from Collins 1992, with permission).

Figure 5.5 Index of heterogeneity (D_s) and bird species diversity (BSD) for shrublands and forests in Texas, Illinois and Delaware (from Roth 1976, with permission).

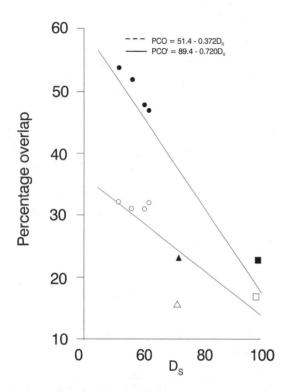

Figure 5.6 Index of heterogeneity (D_s) and percentage overlap (PCO and PCO′) for Illinois forest edge (squares), Texas bushlands (circles), and UD woods (triangles) (from Roth 1976, with permission).

1986). If prey are patchily distributed and the predators tend to concentrate in patches of high prey density, the predator population flourishes, whereas areas with low prey density function temporarily as refuges (May 1978, quoted by Dempster).

5.2.6 Foraging efficiency and heterogeneity

In large patches animals can spend less time searching than in small patches. In fact, the distance between patches varies linearly with the linear dimension of the patch, whereas hunting activity in the patch varies as the square. Larger patches are used in a more specialized way than small patches (MacArthur and Pianka 1966).

In large herbivores such as elk and bison, foraging in a sagebrush–grassland landscape in northern Yellowstone National Park, Wallace *et al.* (1995) found, during the winter time, a response to hetero-

geneity only at a broad scale (landscape), whereas at fine scale (30×30 m) the selection of patches was random. These results indicate that these species move from one area to another according to the abundance of biomass, but that locally they move randomly. This behaviour indicates that when these animals are inside a patch of foraging vegetation no finer-scale choices are made, and that heterogeneity at a small scale is not appreciated. This probably allows a saving in travel energy by moving randomly at a finer scale, but moving non-randomly at broader scale, according to biomass concentration.

Experiments conducted by Gross *et al.* (1995) on bighorn sheep in an artificial enclosure demonstrated a different mechanism: that at a small scale in time and space this species moves directly from one plant to the next closest plant (75% of all moves); 90% of all moves were direct to the three closest plants, and 75% of the time was spent moving directly from one plant to another.

Although a hierarchy in the choice of patch has been suggested by many ecologists in cases where plants are visible, the closest-neighbouring choice was relevant. So, is the response of foragers to a larger scale the sum of small-scale decisions? This is an intriguing point to be analysed. Heterogeneity can probably play a relevant role in the determination of a more efficient food intake strategy, especially in periods of food shortage (Fig. 5.7).

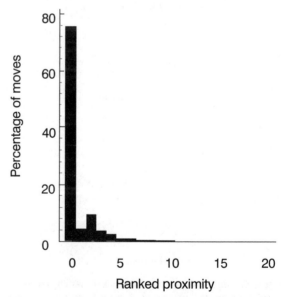

Figure 5.7 Percentage of moves according to plants ranked by their proximity to animals (bighorn sheep) (from Gross *et al.* 1995, with permission).

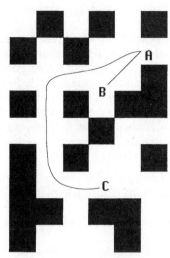

Figure 5.8 In heterogeneous landscapes movement length and complexity are conditioned by the spatial arrangement of patches (from Johnson *et al.* 1992, with permission).

In a heterogeneous landscape animals do not move in a straight line but are strongly conditioned by the spatial arrangement of suitable–unsuitable patches (Johnson *et al.* 1992) (Fig. 5.8).

An example of apparent adaptation to a heterogeneous landscape is presented by Root and Kareiva (1984), studying the movement of cabbage butterflies (*Pieris rapae*). This species places more eggs on isolated hosts than in dense stands, and this behaviour is discussed in terms of a risk-spreading hypothesis.

According to Plowright and Galen (1985), the pollinator flight of bumble bees foraging routes between plants of *Hieracium aurantiacum* is strongly influenced by the heterogeneity of the landscape. In a more uniform landscape the bees fly further than in a stand with landmarks. In areas in which landmarks have been removed the number of backward moves was less than in areas with abundant landmarks; also the flight is more direct and the interplant flight distance significantly longer in uniform areas. Although it is not clear why this behaviour is maintained, bumble bees, like other pollinator insects, are sensitive to the visual landscape and to heterogeneity.

Southwick and Buchmann (1995) found that the homing capacities of honey bees are improved by heterogenous mountain landscapes and decreased in a flat landscape. Honey bees use horizon landmarks to navigate over long distances, and the success of homing decreased with the increase in dis-

tance from the nest. In a mountain landscape honey bees are capable of homing from 9 km, but in flat lands 5 km was the maximum distance from which this species returned (Fig. 5.9).

5.2.7 Heterogeneity and resource use by migrant birds

Forest gaps, important sources of forest heterogeneity, play an important role in oriented patch use for

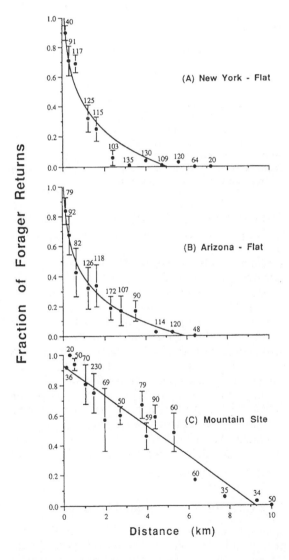

Figure 5.9 Percentage of recaptured foraging bees according to three different areas and different distances (from Southwick and Buchmann 1995, with permission).

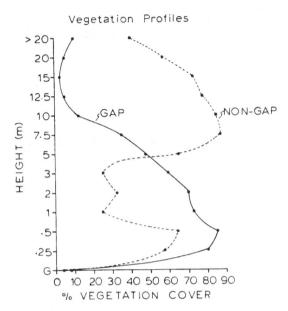

Figure 5.10 Vegetation profile of gap and non-gap areas. Gap areas are characterized by the vegetation cover of the lower layers (from Martin and Karr 1986, with permission).

migratory birds (Martin and Karr 1986). In gaps the foliage profile is significantly different from that in gapless sites (Fig. 5.10).

Other results have been found: birds that use resources concentrated in gaps use more gaps than gapless sites, although their use of gap sites is not restricted to food availability but also applies to other factors, such as perching site, and ultimately the abundance of vegetation close to the soil.

A gap can attract birds because there is more light. If the resources are abundant birds can stay longer and can accumulate at that site. Mixed flocks can behave as a centre of information for other individuals, especially for migrants that have no previous information on resource location.

Finally, Martin and Karr found that bird distribution is consistent with resource abundance when resources are patchy.

5.2.8 Quantification of spatial heterogeneity

Spatial heterogeneity can be defined as the complexity and variability of a system in space (Li and Reynolds 1994), where system properties can be soil nutrients, patch mosaics, plant biomass, animal distribution etc. Variations in spatial heterogeneity

reflect the rate of change in functions and processes.

Li and Reynolds (1994) argued that it is important to produce a clear definition of heterogeneity in order to approach the subject with a good quantitative tool. These authors tested four indices to quantify spatial heterogeneity in simulated landscape maps according to five components of spatial heterogeneity (number of patch types, proportion of patch types, spatial arrangement, patch shape, neighbouring contrast):

1. Fractal dimension. This index measures the complexity of edges (Burrough 1986).
2. Contagion. This index measures the extent to which patches are aggregated (O'Neill *et al.* 1988).
3. Evenness. This index is sensitive to the number of patch types and their proportion in the landscape (Romme 1982).
4. Patchiness. This index measures the contrast of neighbouring patch types in a landscape mosaic (Romme 1982).

Figure 5.11 illustrates the responses of these four indices to the four components of heterogeneity selected (proportion, spatial arrangement, patch shape and neighbouring contrast).

The four indices are to some extent correlated, but also show some redundancy. In particular contagion and evenness are highly correlated, but the correlation between fractal dimension and these indices is weak. This means that the fractal dimension is recommended to be used in combination with the other indices. For more details see Chapter 8.

5.2.9 Summary

- Heterogeneity is the main characteristic of every landscape, and may be defined as the uneven, non-random distribution of objects and is perceived at any scale of investigation.
- Heterogeneity may be considered according to spatial, temporal and functional components.
- Heterogeneity may initiate or exaggerate biological interactions with the environment.
- Heterogeneity assumes a relevant but also contradictory role in the disturbance regime, enhancing or reducing the spread.
- Animals perceive heterogeneity and are also modifiers or producers of heterogeneity. For instance, in ungulates heterogeneity affects grazing efficiency and movements. In invertebrates

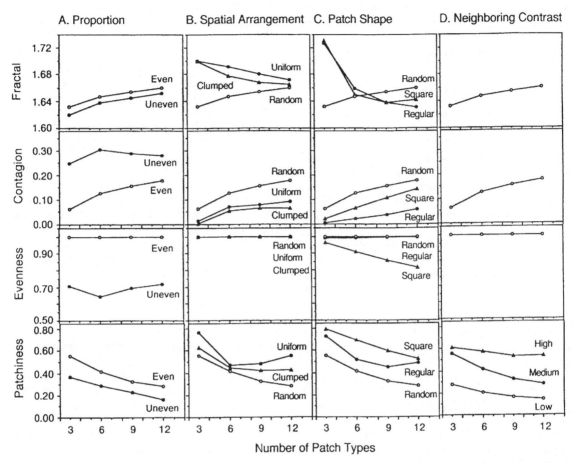

Figure 5.11 Responses of four indices of spatial heterogeneity (fractal, contagion, evenness, patchiness) to the four components of spatial heterogeneity (proportion (uneven, even), spatial arrangement (uniform, clumped, random), patch shape (random, square, regular) and neighbouring contrast (high, medium, low)), using simulated landscape maps (from Li and Reynolds 1994, with permission).

such as insect pollinators heterogeneity affects the food search path and the spatial patterns of foraging flights.

- Heterogeneity may be measured using different indices, such as fractal dimension, contagion, evenness and patchiness.

5.3 ECOTONES

5.3.1 Introduction

The heterogeneous nature of the landscape and the influence of the spatial arrangement of its component patches on many ecosystemic processes have

been recognized (Pickett and White 1985, Hansen *et al.* 1992a,b).

Ecotones were described by Clements (1905) as tension zones where the principal species from adjacent communities meet their limits. Later, Odum (1959) again stressed the importance of defining them as transition zones between two communities.

Ecotones are situated where the rate and dimension of ecological transfers (solar energy, nutrient exchange) change abruptly compared to behaviour in the interior of the patches.

An ecotonal community has species common to both the communities that overlap, and organisms typical of the edges. The tendency to have a high number of species has been called the edge effect. The ecotone may be considered a true habitat, but also a frontier habitat (Ricklefs 1973), where overlap encourages species diversity (Harris 1988).

5.3.2 The importance of ecotones

Ecotones are key structures for the functioning of the landscape: nutrients, water, spore, seeds and animals move across these structures, and play an important role as indicators of climatic change caused by increases in CO_2 and air pollution.

The high biological diversity present in the ecotones, their contribution to system integrity and the high rate of primary and secondary production are other points.

Many data on ecotones are available from different parts of the planet, but are insufficiently well organized to permit reliable comparison; consequently we need more experimental data and affordable models.

In human-modified ecosystems ecotones are the sites with high environmental diversity (woodlot, edges, tree belts) and function as refuges for rare or human disturbance regime-sensitive species.

Long rivers, lakes and coastal lagoon ecotones control the flux of water and nutrients that move from terrestrial to aquatic ecosystems and vice versa.

Many authors have recognized the importance of studying ecotones (Risser 1995, Holland 1988, Naiman et al. 1988, Holland et al. 1991), and important fora have been organized by UNESCO, the Scientific Committee on Problems of the Environment (SCOPE) and by the Council of Scientific Unions (ICSU), to discuss the ecology and management of ecotones. At least five significant publications have been produced (Di Castri et al. 1988, Naiman and Decamps 1990, Holland et al. 1991, Hansen and Di Castri 1992, Risser 1995).

5.3.3 Concepts and definitions

Many authors considered edges as important sites for the study of natural communities (Clements 1897, Livingston 1903, Griggs 1914). Clements (1905) was the first ecologist to introduce the word ecotone, from the Greek *Oikos* (household) and *Tonos* (tension). Later Shelford (1913) and Leopold (1933) confirmed the observation that in ecotones species richness was greater. Since that time the majority of scientific writers have stressed the importance of ecotones (e.g. Weaver and Clements 1928, Odum 1959, Daubenmire 1968, Ricklefs

1973), but only recently have specific publications been devoted to this subject, especially with the increasing interest in landscape ecology (Di Castri et al. 1988, Naiman and Decamps 1990, Holland et al. 1991, Hansen and Di Castri, 1992, Risser 1995). Recently the Ecological Society of America devoted an issue of *Ecological Applications* (Risser 1993) to ecotones and their scaled properties, although the first textbooks of landscape ecology reported on this subject (Forman and Godron 1986, Farina 1993, 1995, Forman 1995).

Ecotones are transitional zones between different habitats, exist at all spatial and temporal scales (Delcourt and Delcourt 1992, Rusek 1992), and are created and maintained by a hierarchy of tension factors, such as air mass dynamics, megatopography, local geomorphology, disturbance, competition, and plant growth and development (Table 5.1). They may be considered as tension areas in which two organizations meet and exchange components, or where the genetic stresses function.

Often ecotones are functional in type, and particularly important are the areas separating subsystems with different degrees of maturity (Margalef 1968).

Among the different possibilities for ecosystems studies are focusing attention on the ecotones, where the high-level exchanges of energy and materials represent a metasystemic approach typical of landscape ecology (Wiens et al. 1985). The diversity and abundance of species (Noss 1983), the flux and accumulation of material and energy (Ranney et al. 1981) and the disturbance diffusion (Pickett and White 1985) are strongly influenced by the borders of the land mosaics. So, for a landscape to function properly it needs ecotones, giving these structures a central role in our understanding of ecological processes (Hansen et al. 1992a,b).

In terrestrial ecosystems the ecotones are seen as soil or vegetation discontinuities, and in a heterogeneous system ecotones are the borders of patches forming the land mosaic (Fig. 5.12).

The environmental conditions can change abruptly, for example between a field and a wood, or between riparian vegetation and the desert, or in a more gradual way from a forest to a mountain prairie by crossing an intermediate shrub cover.

Ecotones can be considered as cellular membranes, functioning as filters and ensuring active and passive transportation according to energy flux or type of material (Forman and Moore 1992). The presence of ecotones in an environmental system is of fundamental importance for the functioning of the entire system. Traditional ecology focused on

Table 5.1 Definitions, functions, time and space scale and factors shaping ecotones

Definitions
1. Sites in which energy exchange and material are highest
2. Transition zones between different habitats
3. Tension zones between systems at different maturities

Functions (ecotones as cellular membranes)
1. Passive diffusion
2. Active diffusion
3. Filter or barrier
4. Accumulation
5. Sink
6. Source
7. Habitat

Spatiotemporal scales and ecotones
Spatial scale
   1. Micro-ecotone
   2. Meso-ecotone
   3. Macro-ecotone
   4. Mega-ecotone
Temporal scale
   1. Ephemeral
   2. Seasonal
   3. Permanent

May be produced by
Exogenous factors
   1. Topography
   2. Climate
   3. Hydrography
Endogenous factors
   1. Ecological succession/competition
   2. Disturbance
   3. Stress
   4. Human activity

Figure 5.12 Ecotone at the tree line (Mt. Cavalbianco, northern Apennines). The beech forest ends sharply at the mountain prairies. The extent of the structural ecotone is a few metres. Human influence is evident in shaping this landscape. A line of old trees was maintained as a border between beech coppice and prairie.

5.3.4 Difficulty of studying ecotones

Owing to their temporary nature and scaling properties ecotones are difficult components of a landscape to investigate (Martinez and Fuentes 1993). Because they are scale dependent their distinctive patterns vanish when closely observed (Margalef 1968). In fact, in many cases the structure and functions of ecotones are not related to physical patterns, especially when the ecotones are created by human disturbance.

The presence of an ecotone is species specific, and the character of an ecotone is not absolute but relative to that species' perception of its surroundings (Shugart 1990).

The localization and the size of an ecotone are two debatable points: generally an edge is considered to be where the contrast between patches is greatest. Using this approach, Turner *et al.* (1990) have studied the variations in annual and perennial vegetation, ants, lizards, birds and mammals along a transect 2700 m long, sampling every 30 m and comparing pairwise sampling, aggregating the samples until only a couple are left (Fig. 5.14).

5.3.5 Spatiotemporal scales and hierarchy of ecotones

The complexity of an ecological system can be observed using a hierarchical approach in which every system is composed of a set of nested subsys-

the interior of homogeneous patches in order to reduce the internal variability and the external influences, whereas landscape ecology, which considers the globality of a system, has to face structurally or functionally defined entities that cause discontinuity of the system (Fig. 5.13).

Since their first definition of a transitional zone between communities, ecotones have been defined as a 'zone of transition between adjacent ecological systems, having a set of characteristics uniquely defined by space and time scales and by the strength of the interactions between adjacent ecological systems' (Di Castri *et al.* 1988).

(a)

(b)

Figure 5.13 (a) Woodland ecotone between *Quercus ilex* (darker) and *Quercus pubescens* (lighter) The ecotone appears as a mixture of the two species, although it is probable that *Quercus pubscens* moves into the *Quercus ilex* stand. The edge is very smooth. **(b)** The ecotone after image classification and cartographic modeling (Mt. Tergagliana, northern Apennines).

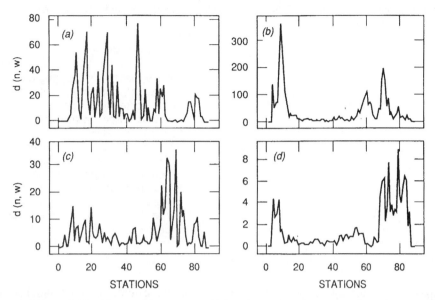

Figure 5.14 Ecotones intersected by (a) ants, (b) mammals, (c) lizards and (d) birds in a grassland. The peaks of differences between the samples indicate discontinuities in the resources or habitats; the peaks are species specific (from Turner *et al.* 1990, with permission).

tems (Allen and Starr 1982, O'Neill *et al.* 1986, Allen and Hoekstra 1992), distinguished by different spatiotemporal scales (Delcourt *et al.* 1983, Delcourt and Delcourt 1987, 1988).

Ecotones representing the borders of different ecological systems exist at all scales and, like ecological systems, can be classified as mega-, macro-, meso- and microecotones. Gosz (1993) distinguished five hierarchical levels for ecotones: biome, landscape, patch, population and individual (Table 5.2). The highest level of this classification is the biome ecotone, which is formed by the blending of patches of different shapes and sizes belonging two adjacent biomes (Fig. 5.15).

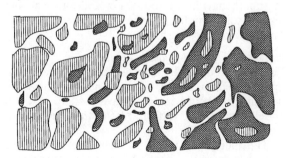

Figure 5.15 Biome ecotone created by contact between two biomes (from Gosz 1993, with permission).

The second level is represented by the environmental mosaic or transition zone, in the shape and size of patches in every biome. The third level is the transition zone between patches composing a biome. The fourth level is represented by the transition between populations. This ecotone exists when a species has a patchy distribution. The fifth level is represented by the ecotone created by an individual (plant) forming a zone of transition due to the effects of competition for nutrients, water and light.

The number of variables that create the ecotonal gradient increases as a proportion of the finest level of scale (Fig. 5.16).

If we consider the temporal scale, we can see that at the scale of 10^4 years ecotones are created by vegetation shift according to climatic changes. At a scale of 10^3 years ecotones are created by the effects of different disturbance regimes of replacing civilizations. At a scale of 100 years ecotones are created by coastal and river network dynamics.

At an annual scale, ecotones are created by flooding regimes. At a seasonal scale ecotones are produced by a climatic event such as snow melting and the availability of water in the soil. At a daily scale ecotones (ephemeral) can be created by thermal constraints in the soil (Fig. 5.17).

Table 5.2 Hierarchical organization of ecotones and variables that create the gradient or constraint on which ecotones exist

Ecotone hierarchy	Probable constraints
Biome	Climate (weather) × topography
Landscape	Weather × topography × soil characteristics
Patch	Soil characteristics × biological vectors × species interaction × microtopography × microclimatology
Population	Interspecies interactions × intraspecies interactions × physiological controls × population genetics × microtopography × microclimatology
Plant	Interspecies interactions × intraspecies interactions × physiological controls × plant genetics × microclimatology × soil chemistry × soil fauna × soil microflora etc.

A landscape can be homogeneous at a scale, without ecotones, and heterogeneous at another scale, with ecotones (Meentemeyer and Box 1987). Risser (1987) introduced a number of principles important for understanding the functioning of an ecological system, related to the presence and functioning of the ecotones:

1. The relationship between structures and processes is not limited to a unique spatiotemporal scale.
2. The importance of a process is scale dependent. For instance, a biogeographic process has a negligible effect on local patterns but is important at a broad scale. An example is local extinction compared to the geographical range of a species.
3. Every group of plants and animals is connected with the environment at a species-specific scale. Every species has its own perception of the environment.
4. The scale of the ecological system is determined by the goal of the research. Some structures and processes are not perceived if the resolution of the investigation is coarser (O'Neill *et al.* 1986).

5.3.6 Ecotone classification

Ecotones exist at all scales and attempts to classify them could seem artificial, but for ecotones recognizable at human scale it is important from a management perspective.

Ecotones may be created by natural or human-induced interactions. Holland (1988) shows a scheme of classification:

1. Ecotones created and maintained by human disturbance regimes (e.g. shelter belts);
2. Ecotones created and maintained by natural processes (e.g. the flooded areas caused by beaver digging);
3. Ecotones produced by natural processes and maintained by human activity (e.g. a strip of riparian forest maintained by man);
4. Ecotones created by human activity and maintained by natural processes (e.g. a flooded area around an artificial reservoir).

Horizontal and vertical ecotones
Most of the information available on ecotones concerns the spatial arrangement of the patches. The vertical ecotones are also implicitly considered important. Since 1961 (MacArthur and MacArthur 1961) bird diversity has been compared with the complexity of the vertical structure of the vegetation. A typical vertical ecotone is represented by the thermic behaviour of soil, water and air mass, but also by soil humidity and the turbulence of gases in the troposphere (Fig. 5.18).

Natural vs. human-induced ecotones
Ecotones produced by natural processes have a soft gradient (Hobbs 1986), whereas those produced by human disturbance regimes have a sharp gradient and the transition zone is often structurally non-existent (Fig. 5.19).

The natural ecotones are sensitive to climatic change and can be profitably used as monitoring areas; the human-induced ecotones can be used as indicators only indirectly for the influence of climatic change on human activity.

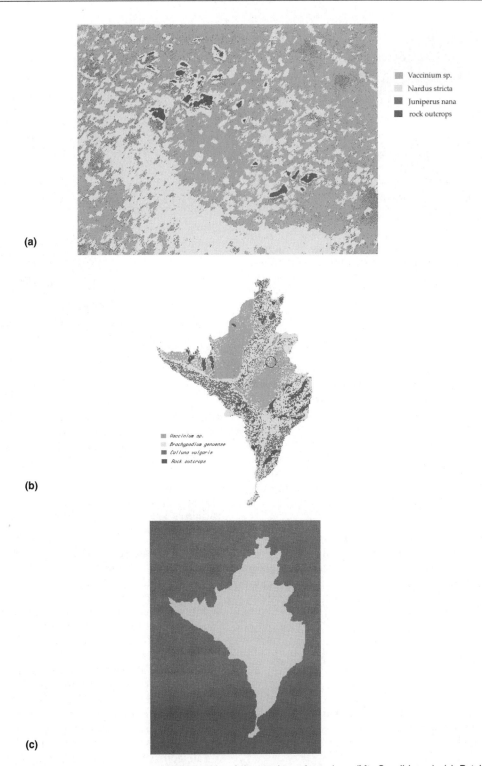

Figure 5.16 Ecotones across scales in a mountain prairie of the northern Apennines (Mt. Cavalbianco). (**a**) Patch ecotone (*Vaccinium* type: *Nardus stricta* type); (**b**) Meso-landscape ecotone. This ecotone exists between slopes (moorland:grassland:rock outcrops:short grasses); (**c**) moorland–beech forest ecotone (from Vannucci 1996, with permission).

Figure 5.17 Ecotones may have an ephemeral life. In this example (Mt. Cavalbianco, northern Apennines) the ecotone has been created by snow accumulation in the mountain prairies. This ecotone is utilized by some micromammals (e.g. snow vole, *Pitymys* spp.) as a winter refuge. Snow cover, because of the low resistance to be crossed, offers a refuge and the possibility of searching for food on the soil surface, escaping predators until the late spring.

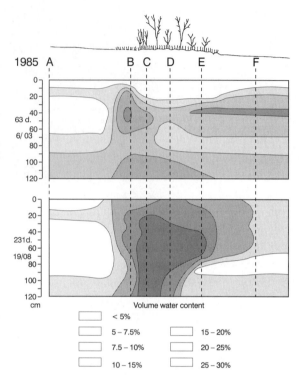

Figure 5.18 Vertical ecotone in a desert vegetation stripe. Soil moisture at two different dates (6 March, showing peak of the dry season, and 16 August after a 50.2 mm rainfall (from Cornet *et al.* 1992, with permission).

Human activity produces an alteration in the spatiotemporal scale of natural processes. In such a way logging produces the same effect as natural forest gaps by treefall, but across mechanisms that are moving at different scales. In other cases human activity has produced modifications and perturbations to natural systems that cannot be observed in natural systems, such as urban cover and other infrastructures, e.g. paved roads and railways.

Genotones
The concept of ecotone can be extended to the spatial arrangement of genes in a population. A good example is the east–west distribution of three genotypes of *Drosophila pseudoobscura* in the southern areas of the southwestern United States (Dobzhansky *et al.* 1977) (Fig. 5.20).

5.3.7 Structural and functional character of ecotones

To understand better the structure, the dynamics and the functioning of ecotones it is necessary to analyse their formation, maintenance, and the inherent and extrinsic factors that ensure their functioning (Fig. 5.21).

The variables that take part in the formation and maintenance of an ecotone can be distinguished into structural and functional types. The structural variables are size, shape, biological structure and structural constraint:

- **Size** The surface or the volume of the ecotone with respect to the size of neighbouring ecological systems and the spatial scale of the fluxes between the systems.
- **Shape** Linear, circular, convoluted etc. This variable seems increasingly relevant to determining the rate of transfer of information, energy and material across ecotones.
- **Biological structure** The distribution of biomass or density of dominant organisms.
- **Structural constraint** The amount of difference between the biological structure of the ecotone and the adjacent ecological systems.
- **Internal heterogeneity** Variance of the changing rate across a discontinuity.
- **Ecotone density** The length of ecotone per unit of land mosaic.

Figure 5.19 Two examples of ecotones. Top: manmade ecotone between non-managed field and mowed pasture. The ecotone is narrow. Bottom: ecotone between *Nardus stricta* and *Vaccinium* spp. on the Apennines. This natural ecotone has an evident gradient.

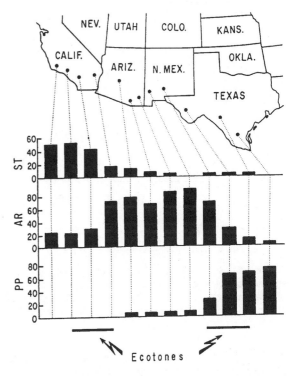

Figure 5.20 Frequency of the spatial distribution in third chromosomes of *Drosophila pseudoobscura* in the southwestern United States. The genotones can be localized where the gradient in chromosome frequencies is steep (from Dobzhansky *et al.* 1977, with permission).

- **Fractal dimension** Rate of complexity of the ecotone shape across a land mosaic.
- **Patch diversity** Richness and evenness of patch types in the land mosaic.
- **Mean patch size** Mean size of patches in a land mosaic.

The functional variables are stability, resilience, quantity of energy and functional contrast:

- **Stability** The degree to which an ecotone resists change when affected by stress.
- **Resilience** The degree to which the ecotone returns to its initial condition after stress.
- **Energetics** The productivity of dominant organisms, the flux of material and energy between the ecotones and the surrounding ecosystems.
- **Functional contrast** The extent of the differences in functional variables between the ecotones and the neighbouring ecosystems.

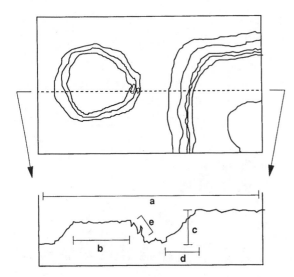

Figure 5.21 Some relevant characters of ecotones: (**a**) density (number of ecotones per unit distance or area; (**b**) width of adjacent patch; (**c**) contrast; (**d**) ecotone width; (**e**) internal heterogeneity (from Hansen *et al.* 1992b, with permission).

- **Porosity** The capacity of an ecotone to change the rate or direction of an ecological flux.

5.3.8 External controls to creation and maintenance of ecotones

In natural systems an edge can be created by external mechanisms that control the ecological systems, or by internal discontinuities that act in the same system.

The environmental response to the changes along a gradient may be gradual or linear and can show an abrupt interruption, probably due to a response with a threshold of one of the components composing the ecological system. Abrupt change along the gradient of a system may have its origin at different levels of organization of the system.

A sharp gradient may occur when a species reaches its tolerance limit, for example, as a response to environmental variables such as temperature, salinity, or pollutants in the soil. In other conditions the abrupt change can happen in the biological response between species, owing to changes in competition. The response may be more complex, depending on the history and evolution of the system, and on the hysteretic relationship between the response and control variables.

Gradual responses to environmental gradients are very common in nature. For instance, the succession from open grassland to forest crosses many intermediate shrub-dominated stages (Fig. 5.22).

5.3.9 Intrinsic controls in the creation and maintenance of ecotones

Edaphic, microclimatic and external disturbance, including human disturbance, affects or creates ecotones. There are also internal factors that can contribute to their maintenance (Odum 1990). Some of these are related to species-specific behaviour. Some species have the capacity to exclude other species, by creating a hostile environment. Mosses have the capacity to modify the pH of the interstitial water, thereby preventing colonization by other plants and creating a long-duration steady state. In fact, the high acidity of water that is in contact with mosses (pH 3–6) prevents the growth of other plants in general.

An ecotone can be maintained by the reproductive aggressiveness of some plants which, thanks to a dense crown, thick root system or high biomass, prevent the seeding of other plants. *Typha* spp. and *Phragmites communis* have the ability to dominate for long periods in marshlands, as does *Nardus stricta* in the mountain prairies. Some alien plants, such as *Helianthus tuberosus,* invade the riverbed in northern Italy, creating a dense monospecific cover and preventing the settlement of native vegetation such as poplar, alder and willow.

Some plants have the capacity to capture sediment transported by surface water or wind. These sediments, transformed by microbial processes, affect the cycle of many nutrients.

Grazing, breeding and digging and seed predation are the main activities of animals in ecotones (Fig. 5.23); for instance, the cutting of trees by beavers creates ecotones composed of shrubs (Johnston and Naiman 1987). The snow vole (*Microtus nivalis*), especially in winter, eats blueberry plants (*Vaccinium* spp.), favouring the growth of grasses in spring (Farina *et al.* 1986). Moles and ants modify the cover of the upland prairies by soil digging and/or the accumulation of vegetal debris in the nests. Termites modify the soil chemistry and wild horses modify vegetation by grazing and the microtopography by trampling.

Although internal factors are important for creating and structuring ecotones, often external events are fundamental to the process. For instance, an ecotone created by a fire depends largely on the fre-

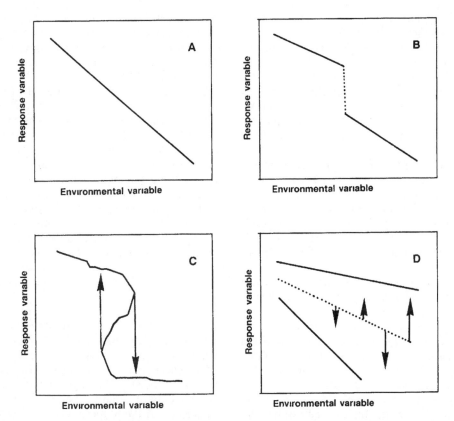

Figure 5.22 Reaction of an ecological system to an environmental variable: (**A**) gradual; (**B**) discontinuous; (**C**) hysteretic; (**D**) multiple responses (from Shugart 1990, with permission).

Figure 5.23 The digging activity of the wild boar creates new ecotones in the upland grassland along the Mediterranean mountain range. The grass cover is broken and new plants can avoid the fierce competition from grass (Logarghena prairies, northern Apennines).

quency of fires, but fires are a function of the regional climate. The same may be observed for the pH changes caused by moss colonization, but moss colonization is itself determined by the hydrological process and a microclimate that favours moss settlement. Internal factors are consequently subordinate to external factors in ecotone creation, structuring and dynamics.

5.3.10. Character of the ecotone

Permeability and vector diffusion
The permeability of ecotones is the capacity to deflect the movement of a vector; it is an inherent character of the edges and is species specific, when animal vectors are considered (Wiens *et al.* 1985).

An ecotone has different permeabilities according to the strength of the vector, such as wind or water, to cross and to transport, and the transported material is more likely to be captured in the ecotone patch when the kinetic fall is higher.

In animals the larger the body mass of a species the more permeable the ecotone (Wiens *et al.* 1985). Physiology may be important in considering the permeability of an ecotone. Species sensitive to microclimate gradients may recognize an ecotone as a barrier, and species with defensive mechanisms can move more easily across ecotones than those with less defensive mechanisms.

In a system dominated by edaphic components the diffusion of water, energy and nutrients varies according to the texture, structure and organic content of the soil. The non-uniform distribution of these elements creates spatial gradients. The patch edges may be sharp, diffuse, linear or convoluted, and can incorporate small or large patches (Stamp *et al.* 1987, Wiens 1992).

The flux of organisms across the edges of the patches is regulated by abiotic, biotic, species-specific and individual factors.

The diffusion of an organism in a homogeneous environment is the same in all directions, but if a gradient such as light, salinity, humidity or structural complexity of the vegetation exists, then the diffusion is more directional (Fig. 5.24).

All these factors may change between the different types of patches. Some differences can be perceived by animals such as birds, which are attracted by a higher diversity and complexity of the vegetation. In fact, at the edges there are more possibilities for nesting and better food resources. Predation may also be higher, and in this case edges can be true ecological traps.

Ecotones are important not only for the movement of species but also for the energy and resources moved by species such as ants, that move material from one patch to another. Beavers modify the hydrological flux by creating temporary dams, and this produces changes in many ecosystem cycles (Johnston and Naiman 1987).

Topography affects the dimension of the fluxes owing to the kinetic energy and the surface/volume ratio of the water mass. The high surface/volume ratio of a dam created by beavers increases the changes at the borders, and the reduction in kinetic energy caused by the dam effect increases the sedimentary capacity of the particles (Johnston and Naiman 1987).

When the contrast between neighbouring patches is very high the edge becomes a true barrier. Human activity has favoured these conditions, which in nature are rare. As true membranes the edges can be permeable to some fluxes but impermeable to others. The edges between the different components of a landscape are consequently very important for the properties of all the systems.

Animal movement across ecotones
Abiotic or biotic vectors actively move energy or materials across the system in non-random way. For example, animal vectors have a complicated behaviour when deciding to cross an ecotone. The permeability of an ecotone depends on both active and passive species-specific diffusion, on perception and on the decision to cross (Wiens *et al.* 1985, Wiens 1992).

Passive diffusion (pd) The rate of diffusion, the viscosity and the heterogeneity of the patches are the factors that affect passive diffusion using wind, water or biological vectors, and may be represented by

$$P(x_1,x_2)_{pd} = \phi(d_i,v_j,h_j)$$

where $P(x_1,x_2)_{pd}$ is the probability of passive movement from position x_1 to position x_2, d_i is the diffusion rate, v_j viscosity of patch j and h_j heterogeneity of patch j.

Active diffusion (at) This variable represents the capacity of an organism to move actively in the

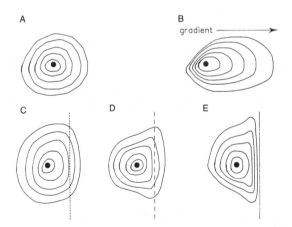

Figure 5.24 Hypothetical behaviour of material or organisms released from a central point (dot): in (**A**) a perfectly homogeneous environment; (**B**) along a gradient; (**C**) a system with a permeable boundary; (**D**) a system with a semipermeable boundary; (**E**) a system with an impermeable boundary (from Wiens 1992, with permission).

environment and depends on the rate of movement, type of movement, density of the organism in the patch, social interactions and habitat preferences.

$$P(x_1,x_2)_{at} = \phi(v_j,h_j) + \phi(r_i,pa_i,d_i,s_i,pr_i)$$

where $P(x_1,x_2)_{at}$ is the probability of movement from point x_1 and point x_2 of the organism i, r_i is the rate of movement of the organism i, pa_i is movement type, d_i organism density, s_i social interactions, and pr_i preferences for different environmental conditions or microhabitats in the patch.

The movement of an organism is influenced by the rate of movement and the shape of the trail. Animals with a high mobility move longer distances than sedentary species, but this varies according to season and site fidelity.

The density of organism in a patch is a very important factor, especially if there are social interactions.

Probability of edge encounter The spatial arrangement, shape and size of an ecotone are important factors in the probability of an animal meeting an ecotone.

$$p(e_k) = \{\phi(a_j,s_j,po_{ij})\} + \{p(x_i,x_2)_{pd} + p(x_1,x_2)_{at}\}$$

where $p(e_k)$ is the probability of encountering an edge k, a_j, s_j is the area and shape of the patch j, po_{ij} the instantaneous position of organism i in the patch j, $p(x_i,x_2)_{pd}$ the probability of passive diffusion, and $p(x_1,x_2)_{at}$ the active diffusion.

Probability of crossing a boundary When an organism is close to a boundary three variables have to be considered: the species-specific permeability, the perception of the border and the level of selection of the patch:

$$p(c_k) = \phi(perm_k,per_{ki},sel_{ji})$$

where $p(c_k)$ is the probability of crossing an ecotone, $perm_k$ is the permeability of the ecotone conditioned by two other variables $perm_k = \phi(s_k,co_k)$, s_k is the sharpness or thickness of the boundary, co_k is the contrast across the boundary between the neighbouring patches, per_{ki} is the perception of boundary k of the organism i, and sel_{ji} is the selection of patch j by organism j.

Cost–benefit balance The permanence of an organism in a patch depends on the balance between costs and benefits of patch occupancy:

$$sel_{ji} = \phi[(c_j/b_j)/(c_m/b_m)]$$

where c_j is the cost associated with the occupancy of patch j, b_j are the benefits associated with the occupancy of patch j and b_m for patch m.

The cost–benefit balance depends on many species-specific variables, such as intra- and interspecific competition, the predation risk, physiological stress, resource availability, the cost of food searching, mating and reproductive success.

$$c_j/b_j = \phi(com_j, pred_j, ps_j, res_j, for_j, m_j, repro_j)$$

where com_j is intra- and interspecific competition, $pred_j$ the predation risk, ps_j physiological stress, res_j resource availability, for_j foraging cost, m_j mating success and $repro_j$ reproductive output.

5.3.11 The function of ecotones in the landscape

Ecotones represent semipermeable membranes across the landscape, modifying the direction, the type and the dimension of material and information exchanged with neighbouring systems (Forman and Moore 1992). For example, Peterjohn and Correll (1984) found that in a small catchment a riverine ecotone can incorporate the surplus of nutrients flowing from the surrounding fields.

Ecotones have been described at several scales, have been found in many environments, and play an active and a passive role in energy and nutrient fluxes. Less understood is their role in the maintenance of landscape stability and the way in which resilience and resistance are transmitted to the adjacent systems (Forman 1981, Baudry 1984, Merriam 1984). For instance, a riparian woodland increases the stability of the neighbouring fields, thereby reducing the effects of river flooding (Fig. 5.25). However, many species of insects harmful to agriculture find in the edges a favourable habitat, reducing the stability of the agroecosystem.

Edges reduce the negative effect of wind, modifying the temperature and soil moisture. Using adaptable plants it should be possible to improve environmental quality and optimize resources. The importance of the ecotone is particularly emphasized in restoration ecology. Ecotones are more easily manipulated than other systems such as forests or grasslands.

5.3.12 The role of ecotones in maintaining local, regional and global diversity

Along edges the abundance and diversity of animals are higher than in the adjacent habitats; this phenom-

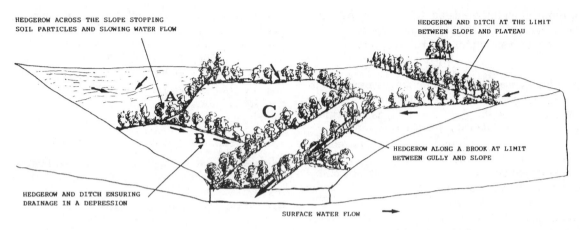

Figure 5.25 Ecotones created in an agricultural mosaic play a fundamental role in preventing erosion, improving the microclimate, and in absorption of pollutants and nutrients (from Burel and Baudry 1990a, with permission).

enon is known as the edge effect. Not all species in an ecotone respond in this way, but the extent and quality of the ecotone are important for biodiversity. Some animals living in ecotones, such as amphibians, spend most of their life in terrestrial habitats but move to ponds to breed. Some species of birds, such as finches, spend the day in ecotones and roost in the forest. Many migratory woodland birds utilize open ecotones as a stopover during migration (Farina 1987, 1988). In many cases it is possible to predict biodiversity according to the density of an ecotone.

The greatest biodiversity is obtained when there is an optimal blend of patches and ecotones. When a landscape is characterized by large patches the number and extension of ecotones are expected to be low. In this landscape biodiversity will also be low. In contrast, when the landscape is highly fragmented it will be the inner species that suffer (Fig. 5.26).

5.3.13 Human impact on the functioning of ecotones

Inner and outer edges are common to natural as well as to manmade landscapes. Some of these edges are obvious, such as the edges between fields and woodlands, or between one type of vegetation and another. Others, such as ecotones of different salinities or pH, are difficult to perceive without tools.

5.3.14 Climatic change and ecotones

There is much evidence that CO_2 released into the atmosphere by factories and cars, and the reduction

of tropical forests, is causing climatic change on our planet. An international programme of the ICSU (International Geosphere–Biosphere Programme, IGBP) was initiated in 1986 to study the physical and chemical processes at the basis of the functioning of the earth system (DeFries and Malone 1989).

In 1987 a report of the UN World Commission on Environment and Development (Brundtland 1987) increased the awareness of the scientific community of the consequences of climatic change for the earth.

The use of ecotones to study climatic change is becoming popular. Suggested for the first time by Griggs (1937), it has recently been used by palaeobotanists to study changes in vegetation cover (Delcourt and Delcourt 1987).

The organisms living in transition zones between two communities may be at the limit of tolerance of local conditions, and will react very soon.

With the increase in temperatures most soils will face drought. Many species will be able to adapt their ecophysiology, but at a broad scale changes in vegetation cover will be expected. At the biome ecotone, for example, a reduction in alpha diversity is expected.

Climatic changes do not have the same effect on all habitats: some have an intrinsic fragility, such as high mountain habitats, and in these cases the climatic change will be more obvious.

5.3.15 The economy and ecotones

The importance of ecotones has been recognized by humans from prehistoric times. The first villages

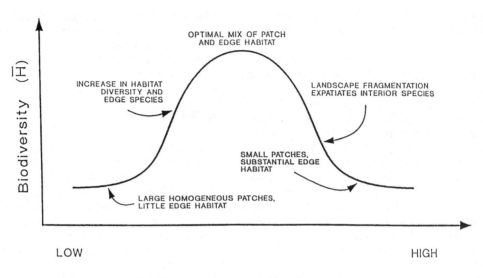

Figure 5.26 Frequency of ecotones and biodiversity value (from Naiman *et al.* 1988, with permission).

and cities were sited on ecotones such as lakes and sea coasts, or in river deltas (Desaigues 1990).

The economic evaluation of ecotones in man-made and modified systems seems of importance. The suppression of many ecotones, such as lagoons and marshes, apparently represents an economic advantage because it makes available new land for cultivation and urban development, but over a long temporal scale it represents a net cost, creating disequilibrium and breaking water and nutrient cycles at landscape level. The productivity and the functionality of a system are assured by ecotones. This is the case with riparian woodland, which represents a buffer zone, reducing freshwater eutrophication (Peterjohn and Correll 1984) and functioning as a barrier to catastrophic flooding.

Humanity has received benefits from ecotones. For example, the hedges separating fields were made to reduce wind effects, changing the microclimate of the soil and favouring plants and animals, but these manmade ecotones are fragile structures, and the recent intensification of agriculture has caused the disappearance of hedges from large parts of rural areas in Europe (Burel and Baudry 1990a,b).

Ecotones like marshes, river deltas and estuaries have been considered too expensive to reclaim; on the other hand, these areas have an invaluable role in maintaining biodiversity.

Using a long-term perspective ecotones are important areas for maintaining a balanced mosaic and are sanctuaries for many species of plants and animals.

5.3.16 Summary

- Ecotones are zones of transition between adjacent ecological systems, having a set of characteristics uniquely defined by space and time scales and by the strength of the interactions between adjacent ecological systems.
- Ecotones are situated where the rate and the dimension of ecological transfers (solar energy, nutrient exchange) have an abrupt change.
- Because of their temporary nature and scaling properties ecotones are difficult structures to investigate, as their distinctive patterns vanish when closely observed.
- Ecotones exist at every scale and may be classified as biome ecotones, landscape ecotones, patch ecotones, population ecotones and individual ecotones.
- Ecotones may be created by natural processes such as flooding and secondary succession, and by human disturbance.
- The structural variables important to ecotones are size, shape, biological structure, structural

constraint, internal heterogeneity, ecotone density, fractal dimension, patch diversity and mean patch size.

- The important functional variables are stability, resilience, energetics, functional contrast and porosity.
- Intrinsic (edaphic conditions, interspecific competition) and extrinsic (human disturbance regime, animal activity) controls are responsible for the creation and maintenance of ecotones.
- Ecotones are selectively permeable to energy and other abiotic and biotic vectors. Animals cross ecotones by passive or active diffusion.
- In human-disturbed landscapes ecotones play a fundamental role in ensuring biological and ecological diversity in the mosaic.
- Ecotones are transition zones in which many organisms live at the limit of tolerance of local conditions, and react very quickly to climatic change. For this reason ecotones are preferred sites for the study of global change and its consequences.
- Ecotones in which human activity has developed greatly at any time are important economic zones.

SUGGESTED READING

Butler, D.R. *Zoogeomorphology. Animals as geomorphic agents.* Cambridge University Press, New York, 1995.

Hansen, A.J. and di Castri, F. (eds.) *Landscape boundaries. Consequences for biotic diversity and ecological flows.* Springer-Verlag, New York, 1992.

Holland, M.M., Risser, P.G., Naiman, R.J. (eds.) *Ecotone. The role of landscape boundaries in the management and restoration of changing environments.* Chapman & Hall, London, 1991.

Kolasa J. and Pickett S.T.A. (eds.) *Ecological heterogeneity.* Springer-Verlag, New York, 1991.

Margalef, R. *Perspectives in ecological theory.* University of Chicago Press, Chicago, 1968.

Naiman, R.J., Holland, M.M., Decamps, H., Risser, P.G. A new UNESCO program: research and management of land:inland water ecotones. *Biology International* 1988, Special Issue 17: 107–136.

Pickett S.T.A. and White P.S. (eds.) *Natural disturbance and patch dynamics.* Academic Press, London, 1985.

Shorrocks B. and Swingland I.R. (eds.) *Living in a patchy environment.* Oxford University Press, New York, 1990.

Turner M.G. (ed.) *Landscape heterogeneity and disturbance.* Springer-Verlag, London, 1987.

REFERENCES

Addicott, J.F., Aho, J.M., Antolin, M.F. *et al.* (1987) Ecological neighborhoods: scaling environmental patterns. *Oikos* **49**: 340–346.

Allen, T.F.H. and Starr, T.B. (1982) *Hierarchy. Perspective for ecological complexity.* Chicago University Press, Chicago, USA.

Allen, T.F.H. and Hoekstra, T.W. (1992) *Toward a unified ecology.* Columbia University Press, New York, USA.

Baudry, J. (1984) Effects of landscape structure on biological communities: the case of hedgerow landscapes. Proc. Int. Sem. IALE *Methodology in landscape ecological research and planning.* Roskilde, Denmark, 15–19 October 1984. Vol. I: 55–65.

Becher, H.H. (1995) On the importance of soil homogeneity when evaluating field trials. *Journal of Agronomy and Crop Science* **74**: 33–40.

Brundtland, G.H. (1987) *Our common future. World Commission on Environment and Development.* Oxford University Press, New York.

Burel, F. and Baudry, J. (1990a) Hedgerows network pattern and processes in France. In: Zonneveld, I.S. and Forman R.T.T. (eds.), *Changing landscapes: an ecological perspective*, Springer-Verlag, Berlin, pp. 99–120.

Burel, F. and Baudry, J. (1990b) Structural dynamic of a hedgerow network landscape in Brittany, France. *Landscape Ecology* **4**: 197–210.

Burrough, P.A. (1986) *Principles of geographical information systems for land resources assessment.* Oxford University Press, Oxford.

Butler, D.R. (1995) *Zoogeomorphology. Animals as geomorphic agents.* Cambridge University Press, New York.

Clements, F.C. (1897) Peculiar zonal formations of the Great Plains. *American Naturalist* **31**: 968.

Clements, F.C. (1905) *Research methods in ecology.* University Publishing Co., Lincoln, Nebraska, USA.

Collins, S.L. (1992) Fire frequency and community heterogeneity in tallgrass prairie vegetation. *Ecology* **73**: 2001–2006.

Cornet, A.F., Montana, C., Delhoume, J.P., Lopez-Portilo, J. (1992) Water flows and the dynamics of desert vegetation stripes. In: Hansen, A.J. and di Castri, F. (eds.), *Landscape boundaries. Consequences for biotic diversity and ecological flows.* Springer Verlag, New York, pp. 327–345.

Daubenmire, R. (1968) *Plant communities.* Harper and Row, New York.

DeFries, R.S. and Malone, T.F. (eds.) (1989) *Global change and our common future: papers from a forum,* National Academy Press, Washington DC.

Delcourt, H.R. and Delcourt, P.A. (1988) Quaternary landscape ecology: relevant scales in space and time. *Landscape Ecology* **2**: 23–44.

Delcourt, P.A. and Delcourt, H.R. (1987) Long-term forest dynamics of temperate forests: applications of paleoecology to issues of global environmental change. *Quaternary Science Review* **6**: 129–146.

Delcourt, P.A. and Delcourt, H.R. (1992) Ecotone dynamics in space and time. In: Hansen, A.J. and di Castri, F. (eds.), *Landscape boundaries. Consequences for biotic diversity and ecological flows.* Springer-Verlag, New York, pp. 19–54.

Delcourt, H.R., Delcourt, P.A., Webb, T. III (1983) Dynamic plant ecology: the spectrum of vegetational change in space and time. *Quaternary Science Review* **1**: 153–175.

Dempster, J.P. and Pollard, E. (1986) Spatial heterogeneity, stochasticity and the detection of density dependence in animal populations. *Oikos* **46**: 413–416.

Desaigues, B. (1990) The socio-economic value of ecotones. In: Naiman, R.J. and Decamps, H. (eds). *The ecology and management of aquatic–terrestrial ecotones.* MAB, UNESCO, Paris, pp. 263–293.

Di Castri, F. , Hansen, A.J., Holland, M.M. (1988) A new look at ecotones: emerging international projects on landscape boundaries. *Biology International,* Special Issue **17**: 1–163.

Dobzhansky, T., Ayala, F.J., Stebbens, G.L., Valentine, J.W. (1977) *Evolution.* Freeman, San Francisco.

Farina, A. (1987) Autumn–winter structure of bird communities in selected habitats of central-north Italy. *Bollettino Zoologia* **54**: 243–249.

Farina, A. (1988) Birds communities structure and dynamism in spring migration in selected habitats of northern Italy. *Bollettino Zoologia* **55**: 327–336

Farina, A. (1993) *L'Ecologia dei sistemi ambientali.* CLEUP, Padova.

Farina, A. (1995) *Ecotoni. Patterns e processi ai margini.* CLEUP, Padova.

Farina, A., Tomaselli, M., Ferrari, C. (1986) Studi sui rapporti tra vegetazione e *Microtus nivalis* (Martins) nell'Appennino reggiano. Prime osservazioni (Mammalia, Rodentia). Atti 51 Conv. UZI, Roma. *Bollettino Zoologia* **53**: 86.

Forman, R.T.T. (1981) *Interaction among landscape elements: a core of landscape ecology.* Proc. Int. Congr. Neth. Soc. Landscape Ecology, Vedhoven 1981, Pudoc, Wageningen, pp. 35–48.

Forman, R.T.T (1995) *Land Mosaics. The ecology of landscapes and regions.* Cambridge University Press, Cambridge.

Forman R.T.T. and Godron M. (1986) *Landscape ecology,* Wiley, New York.

Forman, R.T.T. and Moore, P.N. (1992) Theoretical foundations for understanding boundaries in landscape mosaics. In: Hansen, A.J. and di Castri, F. (eds.), *Landscape boundaries. Consequences for biotic diversity and ecological flows.* Springer-Verlag, New York, pp. 236–258.

Glenn, S.M., Collins, S.L., Gibson, D.J. (1992) Disturbance in tallgrass prairies: local and regional effects on community heterogeneity. *Landscape Ecology* **7**: 243–251.

Gosz, J.R. (1993) Ecotone hierarchies. *Ecological Applications* **3**: 369–376.

Griggs, R.F. (1914) Observations on the behaviour of some species on the edges of their ranges. *Bulletin of the Torrey Botanical Club* **41**: 25–49.

Griggs, R.F. (1937) Timberlines as indicators of climatic trends. *Science* **85**: 251–255.

Gross, J.E., Zank, C., Hobbs, N.T., Spalinger, D.E. (1995) Movement rules for herbivores in spatially heterogeneous environments: responses to small scale pattern. *Landscape Ecology* **10**: 209–217.

Hansen, A.J. and di Castri, F. (eds.) (1992) *Landscape boundaries. Consequences for biotic diversity and ecological flows.* Springer-Verlag, New York.

Hansen, A.J., di Castri, F., Naiman, R.J. (1992a) Ecotones: what and why? In: Hansen, A.J. and di Castri, F. (eds.), *Landscape boundaries. Consequences for biotic diversity and ecological flows.* Springer Verlag, New York, pp. 9–46.

Hansen, A.J., Risser, P.G., di Castri, F. (1992b) Epilogue: biodiversity and ecological flows across ecotones. In: Hansen, A.J. and di Castri, F. (eds.), *Landscape boundaries. Consequences for biotic diversity and ecological flows.* Springer Verlag, New York, pp. 423–438.

Hantush, M.M. and Marino, M.A. (1995) Continuous time stochastic analyses of groundwater flow in heterogeneous aquifers. *Water Resources Research* **31**: 565–575.

Harris, L.D. (1988) Edge effects and conservation of biotic diversity. *Conservation Biology* **2**: 330–332.

Hastings, A. (1990) Spatial heterogeneity and ecological models. *Ecology* **71**: 426–428.

Hobbs, E.R. (1986) Characterizing the boundary between California annual grassland and coastal sage scrub with differential profiles. *Vegetatio* **65**: 115–126.

Holland, M.M. (1988) SCOPE/MAB Technical consultations on landscape boundaries. Report of a SCOPE/MAB workshop ecotones. In: Di Castri, F., Hansen, A.J., Holland, M.M. (eds.), A new look at ecotones: emerging international projects on landscape boundaries. *Biology International*, Special Issue **17**.

Holland, M.M., Risser, P.G., Naiman, R.J. (1991) *Ecotone. The role of landscape boundaries in the management and restoration of changing environments.* Chapman & Hall, London.

Johnson, A.R., Wiens, J.A., Milne, B.T., Crist, T.O. (1992) Animal movements and population dynamics in heterogeneous landscapes. *Landscape Ecology* **7**: 63–75.

Johnston, C.A. and Naiman, R.J. (1987) Boundary dynamics at the aquatic–terrestrial interface: the influence of beaver and geomorphology. *Landscape Ecology* **1**: 47–58.

Kolasa, J. and Pickett, S.T.A (1991) *Ecological heterogeneity.* Springer-Verlag, New York.

Kolasa, J. and Rollo, CD. (1991) Introduction: the heterogeneity of heterogeneity: a glossary. In: Kolasa, J. and Pickett, S.T.A *Ecological heterogeneity.* Springer-Verlag, New York, pp. 1–23.

Leopold, A. (1933) *Game management.* Scriber, New York.

Levin, S.A. (1976) Population dynamic models in heterogeneous environments. *Annual Review of Ecological Systems* **7**: 287–310.

Levins, R. (1968) *Evolution in changing environments.* Princeton University Press, Princeton, NJ.

Li, H. and Reynolds, J.F. (1994) A simulation experiment to quantify spatial heterogeneity in categorical maps. *Ecology* **75**: 2446–2455.

Livingston, B.E. (1903) The distribution of the upland societies of Kent County. *Michigan Botanical Gazette* **35**: 36–55.

MacArthur, R.H. and MacArthur, J.W. (1961) On bird species diversity. *Ecology* **42**: 594–598.

MacArthur, R.H. and Pianka, E.R. (1966) On optimal use of patchy environment. *American Naturalist* **100**: 603–610.

MacArthur, R.H., MacArthur, J.W. and Preer, J. (1962) On bird species diversity. II. Prediction of bird census from habitat measurements. *American Naturalist* **96**: 167–174.

Margalef, R. (1968) *Perspectives in ecological theory.* University of Chicago Press, Chicago.

Martin, T.E. and Karr, J.R. (1986) Patch utilization by migrating birds: resource oriented. *Ornis Scandinavica* **17**: 165–174.

Martinez, E. and Fuentes, E. (1993) Can we extrapolate the California model of grassland–shrubland ecotone? *Ecological Applications* **3(3)**: 417–423.

May, R.M. (1978) Host–parasitoid systems in patchy environments, a phenomenological model. *Journal of Animal Ecology* **47**: 833–843.

Meentemeyer, V. and Box, E.O. (1987) Scale effects in landscape studies. In: Turner, M.G. (ed.), *Landscape heterogeneity and disturbance.* Springer-Verlag, New York, pp. 15–36

Merriam, G. (1984) *Connectivity: a fundamental characteristic of landscape pattern.* Proc. Int. Sem. IALE Methodology in landscape ecological research and planning. Roskilde, Denmark, 15–19 October 1984. Vol. I: 5–15

Naiman, R.J. and Decamps, H. (1990) *The ecology and management of aquatic–terrestrial ecotones.* Parthenon Publishing Group, Park Ridge, NJ.

Naiman, R.J., Holland, M.M. , Decamps, H., Risser, P.G. (1988) A new UNESCO program: research and management of land:inland water ecotones. *Biology International*, Special Issue **17**: 107–136.

Noss, R.F. (1983) A regional landscape approach to maintain diversity. *BioScience* **33**: 700–706.

Odum, E.P. (1959) *Fundamentals of ecology*, 2nd edn. W.B. Saunders Company, Philadelphia.

Odum, W.E. (1990) Internal processes influencing the maintenance of ecotones: do they exist? In: Naiman, R.J. and Decamps, H. (eds.), *The ecology and management of aquatic–terrestrial ecotones.* MAB, UNESCO, Paris, pp. 91–102.

O'Neill, R.V., DeAngelis, D.L., Waide, J.B., Allen, T.F.H. (1986) *A hierarchical concept of ecosystems.* Princeton University Press, Princeton.

O'Neill, R.V., Krummel, J.R., Gardner, R.H. *et al.* (1988) Indices of landscape pattern. *Landscape Ecology* **1**: 153–162.

Peterjohn, W.T. and Correll, D.L. (1984) Nutrients dynamics in an agricultural watershed: observations on the role of a riparian forest. *Ecology* **65**: 1466–1475.

Pickett, S.T.A. and Cadenasso, M.L. (1995) Landscape ecology: spatial heterogeneity in ecological systems. *Science* **269**: 331–334.

Pickett, S.T.A. and White, P.S. (1985) *The ecology of natural disturbance and patch dynamics.* Academic Press, New York.

Plowright, R.C. and Galen, C. (1985) Landmarks or obstacles: the effect of spatial heterogeneity on bumble bee foraging behavior. *Oikos* **44**: 459–464.

Ranney, J.W., Bruner, M.C., Levenson, J.B. (1981) The importance of edge in the structure and dynamics of forest islands. In: Burgess, R.L., Sharpe, D.M. (eds.), *Forest island dynamics in man-dominated landscapes.* Springer-Verlag, New York, pp. 67–95.

Ricklefs, R.E. (1973) *Ecology.* Chiron Press.

Risser, P.G. (1987) Landscape ecology: state-of-art. In: Turner, M.G. (ed.) *Landscape heterogeneity and disturbance.* Springer-Verlag, New York, pp. 3–14.

Risser, P.G. (1993) Ecotones. Ecotones at local to regional scales from around the world. *Ecological Applications* **3**: 367–368.

Risser, P.G. (ed.) (1995) Understanding and managing ecotones. Proceedings of the third International SCOPE/UNEP Workshop on Ecotones, Moscow 1993. *Ecology International* **22**.

Romme, W.H. (1982) Fire and landscape diversity in subalpine forests of Yellowstone National Park. *Ecological Monographs* **52**: 199–221.

Root, R.B. and Kareiva, P.M. (1984) The search for resources by cabbage butterflies (*Pieris rapae*): ecological consequences and adaptive significance of markovian movements in a patchy environment. *Ecology* **65**: 147–165.

Rotenberry, J.T. (1985) The role of habitat in avian community composition: physiognomy or floristic? *Oecologia* **67**: 213–217.

Roth, R. (1976) Spatial heterogeneity and bird species diversity. *Ecology* **57**: 773–782.

Rusek, J. (1992) Distribution and dynamics of soil organisms across ecotones. In: Hansen, A.J. and di Castri, F. (eds.), *Landscape boundaries. Consequences for biotic diversity and ecological flows.* Springer Verlag, New York, pp. 196–214.

Scoones, I. (1995) Exploiting heterogeneity: habitat use by cattle in dryland Zimbabwe. *Journal of Arid Environment* **29**: 221–237.

Shelford, V.E. (1913) *Animal communities in temperate America.* University of Chicago Press, Chicago.

Shorrock, B. and Swingland, I.R. (eds.) (1990) *Living in a patchy environment.* Oxford University Press, Oxford.

Shugart, H.H. (1990) Ecological models and the ecotones. In: Naiman, R.J. and Decamps, H. (eds). *The ecology and management of aquatic–terrestrial ecotones.* MAB, UNESCO, Paris, pp. 23–36.

Southwick, E.E. and Buchmann, S.L. (1995) Effects of horizon land marks on homing success in honey bees. *American Naturalist* **146**: 748–764.

Stamp, J.A., Buechner, M., Krishnan, V.V. (1987) The effects of edge permeability and habitat geometry on emigration from patches of habitat. *American Naturalist* **129**: 533–552.

Turner, M.G. (ed.) (1987) *Landscape heterogeneity and disturbance.* Springer-Verlag, New York.

Turner, S.R., O'Neill, R.V., Conley, W., Conley, M., Humphries, H. (1990) Pattern and scale: statistics for landscape ecology. In: Turner, M.G. and Gardner, R.H. (eds.), *Quantitative methods in landscape ecology: the analysis and interpretation of landscape heterogeneity.* Springer-Verlag, New York, pp. 17–49.

Vannucci, G. (1996) *Studio della truttura spaziale della vegetazione in un ecotone montano appenninico attraverso un approccio multiscalare.* Unpublished thesis.

Wallace, L.L., Turner, M.G., Romme, W.H., O'Neill, R.V., Wu, Y. (1995) Scale of heterogeneity of forage production and winter foraging by elk and bison. *Landscape Ecology* **10**: 75–83.

Weaver, J.E. and Clements, F.C. (1928) *Plant ecology.* McGraw-Hill, New York.

Wiens, J.A. (1992) Ecological flows across landscape boundaries: a conceptual overview. In: Hansen, A.J. and di Castri, F. (eds.), *Landscape boundaries. Consequences for biotic diversity and ecological flows.* Springer-Verlag, New York, pp. 217–235.

Wiens, J.A., Crawford, C.S., Gosz, R. (1985) Boundary dynamics: a conceptual framework for studying landscape ecosystems. *Oikos* **45**: 421–427.

Principles of landscape dynamics

6.1 INTRODUCTION

The land mosaic generally has no permanent shape but changes in quality, size, shape and spatial arrangement of the patches. This dynamic is the result of complex, multiscalar processes and has tremendous importance for most living organisms.

Flows are the linking processes in the mosaic (Wiens 1995) and movements are related to flows. Movements are scaling processes and understanding them requires a multiscale approach. For this purpose fractal analysis seems a very promising tool (Krummel *et al.* 1987, Milne 1991; see also Chapter 8).

The movement of energy, matter and organisms in a heterogeneous landscape mosaic finds constraints or gradients where a patch homogeneous for some characters meets a different patch. Many studies are available on the functioning of small-scale systems (ecosystem approach) but there is little information on large-scale systems, largely because of the relatively new scales adopted and also because of the difficulties of integrating or combining at a higher level the information on subordinate levels. Internal and external stimuli can produce changes in the system. Landscapes, like other natural systems, can behave differently when affected by natural or human-induced stresses. A system can react by resisting the external perturbation, which is measured by the degree to which a variable is changed from its equilibrium.

The dynamic of a landscape depends on at least four major factors:

1. Disturbance frequency;
2. Rate of recovery from disturbance;
3. The size or spatial extent of the disturbance events;
4. The size or spatial extent of the landscape.

Turner *et al.* (1993) have distinguished the temporal parameter (T) as being the ratio of the disturbance interval (i.e. the interval between successive disturbances) to the recovery time (the time required for a disturbed area to regain a 'mature' stage). Three cases are possible:

1. The disturbance interval is longer than the recovery time (T > 1).
2. The disturbance interval is equal to the recovery time (T = 1).
3. The disturbance interval is shorter than the recovery time (T < 1).

The spatial parameter (S) is defined as the ratio between the size of disturbance and the size of the landscape of interest. There are two types of spatial ratio:

1. Disturbance is larger than the landscape of interest; in this case the landscape dynamic cannot be predicted because the landscape is too small to characterize the effect and recover the disturbance.
2. Disturbance is smaller than the extent of landscape.

Many landscapes can be affected by different disturbance regimes acting at different spatiotemporal scales, and it is the result of these regimes and the relative recovery that creates the patchiness.

An understanding of landscape dynamics has tremendous implications for landscape management and reserve planning. It is clear that the larger a reserve the higher the probability that a landscape can incorporate all natural disturbances.

The scale-dependent nature of the concept of landscape equilibrium allows us to accept this concept only for a specific spatial and temporal scale.

The spatial setting of an ecosystem within a landscape influences many of that ecosystem's properties, and this can be well demonstrated, especially when data are available from long-term studies (Kratz *et al.* 1991). However, it is not well understood that a landscape influences the dynamics of its component ecosystems, but there is little doubt that landscape influence is important in controlling ecosystem processes. For this reason the application of the non-equilibrium concept to the study of an ecosystem seems very promising. It is clear that an ecosystem shows variability when an external stimulus (stress) such as acid rain or climatic change occurs. We can expect a low variability in ecosystems in which the landscape constraint is low, and high variability when the position of the ecosystem in the landscape is more exposed to change. For instance, on a mountain ridge the weather conditions can change abruptly in a short time, and we expect strong influences on the ecosystems that occupy this position. But even in ecosystems that receive the same weather we can expect landscape position to be a predictor of temporal and spatial variability. For instance, geological processes such as erosion and deposition can cause land forms that have a profound influence on soil characteristics. These differences may have strong effects on biogeochemical processes such as water retention, chemical buffering capacities and the dynamics of microbial communities. From the study of Kratz *et al.* (1991), in which 68 parameters were measured, despite the implicit limitations of a temporal grain size of one year, with an extent of

five years, it emerged that individual locations in each landscape differed from one other in annual variability. At each site the variability patterns of at least a subset of the parameters were associated with a particular subset of parameters, demonstrating that landscape influences the temporal dynamics of ecosystems in a predictable way.

6.2 STABILITY IN LANDSCAPES

We can consider stability as a family of processes having the capacity to confirm patterns and processes in time. The stability of a landscape could be wrongly conceived if no scaling perspective is used. We can expect stability across a range of spatial and temporal scales, therefore the stability of a landscape must be scaled according to the process that we intend to describe.

Generally, at a broad scale changes in landscape structure and composition require a long period of time to take place, but at a small scale this can happen in a very short time.

In natural landscapes some structuring elements are more fragile than others. For instance, riverine vegetation is conditioned by seasonal flooding that can completely destroy the physical substrate on which plants are rooted, but in old-growth forests, for example, a severe fire may destroy some plants but not all the forest, and the disturbance can easily be incorporated. A severe fire should be a rare event occurring at a scale of a hundred years.

The stability of a tropical forest is connected to the climatic regime by the rainfall. This stability may be reduced by rare events such as large-scale hurricanes or by changes in atmospheric circulation caused by stream jet currents. On the other hand, riverine vegetation in the Mediterranean region is under constant threat of unexpected flooding, which can dramatically change the shape and structure of the substrate.

At the landscape scale generally stability is manifested by complicated patterns that locally present frequent changes, but which at a large scale maintain the same shape. So, at the watershed scale the riparian vegetation appears more stable than when observed along a separate sector of the river channel. Such metastability is a relevant character of every landscape.

Perturbations in stability occur in natural as well as in anthropogenic landscapes, producing apparently the same effects, but caution has to be observed as regards the functioning of the perturbed systems. For instance, human effects on the landscape vary according to the severity with which humans change the landscape and use the resources. The hunting and gathering activity of sparse primitive populations probably had little effect on the landscape structure: this can be compared with the effects of large mammals in African savannas living in undisturbed conditions without area restrictions (Belsky 1995). In this case the mammals are conditioned by topography, soil quality, energy and water availability. It is the landscape structure that drives the dynamics of the species, although trampling, grazing, urine and manure deposition can produce ephemeral local effects.

When these conditions are threatened with area restriction due to human influence the pressure of mammals on the landscape becomes the main shaping force, producing a different landscape in which patches of different quality are created according to differential pressures.

6.3 SELF-ORGANIZING MECHANISMS AND LANDSCAPES

A self-organizing (or self-reinforcing) system may be defined as one in which structures and processes reinforce each other. Such a system can maintain order through internal interactions and can be considered as antichaotic. Unlike a chaotic system, which is highly sensitive to original conditions, in this case the initial conditions are channelled into the same final state.

In ecosystems there is a strong positive feedback between plant mutualists and heterotrophs. A complicated chain of interspecific interactions creates guilds that reduce the probability of the threshold food chain collapsing. These interactive self-reinforcing communities are spatially integrated in a landscape whose dynamics are themselves self-reinforcing.

In some landscapes disturbances are incorporated but in other cases they are magnified; in both cases there is a process of reinforcement of the structures. For instance, large stands of undisturbed old-growth forests are less susceptible to catastrophic fire, thereby acting to preserve forest structure. On the other hand, a fragmented landscape composed of young stands and a grass layer is more susceptible to catastrophic fire. The Mediterranean landscape is a typical example of this: the landscape

is highly susceptible to fire and most plants and their structures favour fire contagion.

6.4 LANDSCAPE SHAPING FACTORS

A hierarchy of factors act to shape a landscape. In general terms the most important factor is the climatic conditions. Recently Bailey (1995) described the ecoregions of the United States, classifying them into dominions, divisions and provinces.

The character of air masses is the first criterion, followed by intensity of radiation and seasonal changes in the climatic regime.

Soil is modified and profoundly influenced by climate. According to the vegetation criterion, vegetation has some capacity to modify soil but the causes and effects are difficult to separate.

Animals represent the last step in this hierarchy, but there are many exceptions, such as in the semi-arid regions in which termite activity can change the functioning of vegetation.

6.5 CHANGES IN HUMAN-PERTURBED LANDSCAPES

Landscapes can be changed by many different natural and human-induced perturbation regimes. Natural perturbations like flooding, fire and thunderstorms have a profound influence on the historical evolution of landscapes, but generally human influence overwhelms or masks these natural processes because of the higher frequency of occurrence.

Relevant landscape-changing processes directly related to human activity are:

- Agricultural intensification;
- Agriculture abandonment;
- Fire suppression;
- Deforestation;
- Livestock grazing;
- Development.

6.5.1 Agricultural intensification

This process generally produces a decrease in landscape mosaic complexity, a simplification of many geochemical cycles, a reduction of many ecological processes, a simplification of the trophic chain and a decrease in system resilience.

The monoculture style of modern agriculture is a very simplified system in which a considerable input of fertilizer allows farmers to maintain a high rate of harvest. In this type of landscape linear features are dominant and generally most of the soil is deeply perturbed by seasonal ploughing.

6.5.2 Agriculture abandonment

Land abandonment is one of the more conspicuous phenomena in developed countries, from Europe to North America. The abandonment of agriculture is a pattern common to all industrialized countries, especially in hilly and mountainous areas. Well known across the Mediterranean basin, it is a relatively recent phenomenon elsewhere in recent decades, and will be discussed further at the end of this chapter.

6.5.3 Fire suppression

Most natural and manmade landscapes have a patchy structure that is important for many species of plants and animals. When a new disturbance regime is introduced into a patchy landscape many changes occur in the system. When the fire regime in a forested landscape is suppressed, a cascade effect can be recognized. Baker (1992) has used a model to study the effect of fire suppression, analysing scenarios since before the settlement of Europeans in the Boundary Canoe, Minnesota, until the present. To assess the landscape changes seven measures of landscape structure were used

- Mean patch size (total number of pixels in the study area/number of patches);
- Mean shape s = (0.282 × perimeter)/area 0.5;
- Mean fractal dimension (Krummel *et al.* 1987);
- Shannon index of diversity H = − $pi\Sigma$log(pi) where pi is the proportion of landscape occupied by patch of age i;
- Mean richness is the mean number of patch age;
- Mean angular second moment is an index of fine-scale texture of the landscape (Haralick *et al.* 1973).

Baker found that when disturbance size and frequency decline because of settlement and suppression, immediate changes occur in the landscape structure as monitored by shape, Shannon diversity and richness. Some changes occur later (age, fractal

dimension) or after hundreds of years (size, angular second moment).

When the disturbance regime is changed from presettlement to settlement some measures react immediately (age, shape, Shannon diversity, richness, second angular moment), but on others the effects are not visible (size, fractal dimensions).

After decades of fire suppression and the recognition of its negative effect on landscape structure, prescribed fire has been utilized in different situations across North America.

6.5.4 Deforestation

Despite an increase in the use of plastics, metals and 'virtual' materials, deforestation is a conspicuous phenomenon in many parts of the world, especially in boreal and tropical forests. Deforestation modifies the structure and functions of the landscape, increases fragmentation and the amount of edge habitat, increases the diversity of stand age, creates linear borders and facilitates the immigration of open-space species into the forest interior.

It is necessary to distinguish between deforestation of primeval forests (tropical and boreal), which

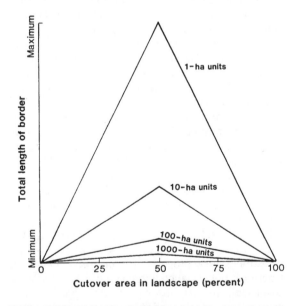

Figure 6.1 Total length of edge between forest and cutover area across different cutting-unit sizes (from Franklin and Forman 1987, with permission).

generally has severe consequences, from coppicing in temperate forests. In this case the logging is periodic and the perturbation regime is incorporated by the woodland.

In primeval boreal forests the spatial arrangement of the logged stands is tremendously important in maintaining biodiversity. Franklin and Forman (1987) proposed a spatial model based on a checkerboard shape (Fig. 6.1), in which cutting is dispersed as evenly as possible.

The landscape is strongly affected by the choice of forest exploitation patterns. The probability of disturbance, e.g. fire and windthrow, means that species diversity is strongly sensitive to the spatial arrangement of the land mosaic.

White and Mladenoff (1994) have studied the evolution of a 9600 hectare landscape in Northern Wisconsin from 1860 until the present. The forest composition, from pre- to post-settlement, changed from a landscape dominated by *Tsuga canadensis* to hardwoods (*Acer saccharum, Betula allegheniensis*). From 1860 to 1931 large disturbance processes associated with logging were dominant. After 1931 succession was the dominant process, with an increase in homogenization of the landscape.

6.5.5 Livestock grazing

Many regions of the world are affected by the spread of grazing disturbance by livestock. The effects of grazing on the structure and function of landscapes are dramatic. Domestic grazers create landscape patterns with different environmental effects according to the conditions in which they are reared. This disturbance is density dependent and plant adaptation to grazing is strictly linked to the severity of disturbance. Often, when the density of grazers is high, other effects are added, such as trampling and an increase in soil nitrogen content. The landscape reacts promptly to grazing regimes, modifying the plant community. Under a moderate grazing regime plant diversity is enhanced, but when the grazing pressure increases a rapid drop in diversity is experienced.

Often, as in most of the US western grasslands, the modification of the plant mosaic has been so severe that shrubby steppes have replaced annual and perennial grasslands.

To different extents mountain prairies across Europe have for many centuries experienced a grazing pressure that has profoundly modified the composition and spatial arrangement of vegetation patch types.

6.5.6 Development

Urbanization and infrastructures like roads, railways and airports cover broad areas of the earth, with dramatic consequences for the landscape. The covering of soil and destruction of natural vegetation are added to the modification of both surface and underground hydrological nets.

A developed area is a tremendous sink demanding and absorbing enormous quantities of energy (from water to oil and food) and is a source for degraded energy (heat), waste water and solid garbage.

The increase in world population and the urbanization of many regions are creating the conditions for larger developed areas such as Mexico City and Los Angeles. In these areas the landscape is completely modified and natural patterns and processes are replaced by artificial structures. The capacity of the landscape to incorporate the disturbance is reduced or erased. For instance, if an entire watershed is occupied by developments the functioning of this watershed will always be pathological. Therefore, the human strategy should be to mimic natural processes, but often this is prevented by high costs, lack of information and knowledge and the demand for land for infrastructures.

6.6 PATTERNS OF LANDSCAPE CHANGE: EXAMPLES

Natural and anthropogenic landscapes have different degrees of fragility, which depends on their capacity to change after disturbance.

The interception of fluxes is fundamental for many processes, and a knowledge of the spatial arrangement of the phenomena is critical for a better understanding of the functioning of the system. Unfortunately, landscape changes are not easily detected at short timescales, and to investigate these changes requires good-quality field information across a long time period, which is not always available.

In human-dominated landscapes the variables involved in change are many and related mostly to socioeconomic processes (Fig. 6.2).

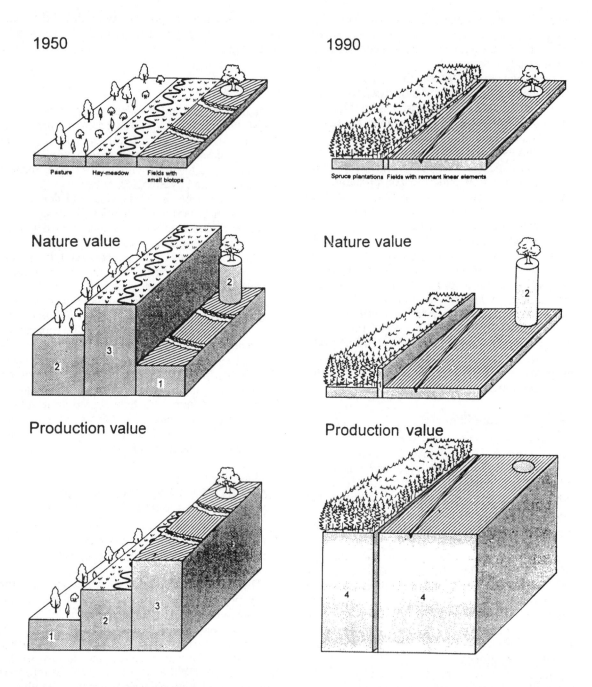

Figure 6.2 Changes in biological value and in production in a Nordic landscape from 1950 to 1990. The land mosaic in 1950 was composed of pastures, hay meadows and fields with hedgerows and scattered trees. In 1990 the mosaic is more simple, composed of spruce plantations and fields with linear thin hedgerows (from Ishe and Norderhaug 1996, with permission).

The uniqueness of landscapes does not permit standard approaches to the study of change. Often this depends on the information available, the microclimatic and topographic characteristics, socioeconomic history and trends.

Simpson *et al.* (1994) investigated changes between 1900 and 1988 in two contiguous Ohio landscapes: a tilled plain landscape and a moraine landscape, which latter has a greater geomorphological diversity and heterogeneity. Using aerial photographs from 1940, 1957, 1971 and 1988, combined with historical archives, the landscape information was transferred to a GIS (geographical information system) for comparison. Different evolutionary patterns were detected: on the moraine agriculture decreased as forest, urban and suburban areas and industrial settlement increased. In the tilled plain agriculture increased until 1988.

Land cover, diversity and evenness were all higher in the moraine than in the plain at all times. The moraine landscape showed more dynamism but the plain showed more inertia. Factors that have influenced the evolution of these landscapes are topography, soil capability and socioeconomic environment, e.g. agriculture policies and patterns of urbanization. Socioeconomic factors have to be coupled to physical and biological factors to understand the dynamics of human-perturbed landscapes.

Skinner (1995), using aerial photographs of a forested watershed of 24 600 ha in northwestern California, compared the spatial structure of vegetation 41 years apart. In this area a fire suppression policy has modified the dynamics of disturbance and the spatial arrangement of patches. Fifty random sample points were selected in two 1:16 000 aerial photographs, one from 1944 and one from 1985. At each sample point were calculated the distance from sample point to the edge of the nearest opening; the distance from the edge of the first opening to the next closest opening; the perimeter of each opening; the area of the opening; the non-opening area around each opening closer than any other non-opening. The opening perimeter was regressed against opening area, and other shape indices were utilized.

The comparison between 1944 and 1985 openings showed a reduction as expected under the fire suppression regime. The distance between openings increased significantly from 1944 to 1985. In conclusion, fire suppression has considerably changed the forest landscape over 41 years, but there were no significant differences in the form or shape of openings.

Changes in landscape composition and their relation to physiographic regions were evaluated by Turner and Ruscher (1988) in Georgia, USA, over a 50-year period (1930–1980). Land use patterns were measured by mean number and size of patches; fractal dimension of patches; edge length between land uses; index of diversity, dominance and contagion (see Chapter 8). The results show clear changes in landscape according to the different physiographic regions (lower coastal plain, upper coastal plain, foothills, mountains) (Fig. 6.3).

The Georgian landscape exhibited the greatest changes in the foothills, and in general appears to be less fragmented at the present time. The complexity of patches decreased as shown in Fig. 6.3 using the fractal index. These changes had a great effect on fauna, favouring forest species and affecting edge species.

The landscape changes that have occurred in the past 160 years have been quantified by Iverson (1988) across Illinois State. Eleven soil attributes were used in the analysis. Naturally derived land types are closely influenced by landscape attributes (44–83% of variance), but urban type, strip mines and quarries and reforested lands are not associated with landscape attributes (17–30%). This study clearly demonstrates the importance of landscape scale in evaluating land changes.

6.7 MEDITERRANEAN LANDSCAPES AS AN EXAMPLE OF PERTURBATION-DEPENDENT HOMEORETHIC SYSTEMS

The Mediterranean landscape has been extensively modified by anthropogenic disturbance over thousands of years. Local ecologically related activities of crop raising and livestock management have been overlapped, time after time, by the intrusion of alien cultures during westbound migrations (Naveh and Vernet 1991). We have little information on the effects of these combined factors, but we can imagine some destructive effects of a local metastability

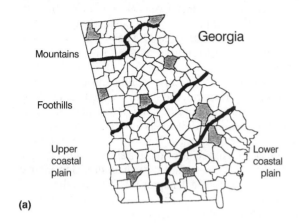

(a)

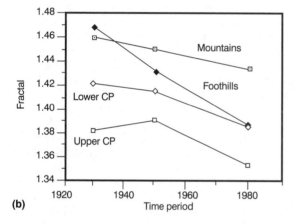

(b)

Figure 6.3 (a) Map of Georgia, counties and physiographic regions. (b) Fractal dimension of patches in each physiographic region using all types of land use (from Turner and Ruscher 1988, with permission).

reached by a fine-grained interaction between residents and their surrounding landscape.

Some models of alien culture, such as the Roman 'centuriazione' (Caravello and Giacomin 1993), seem very successful, for example persisting to the present time in the Po Valley (northern Italy).

The roughness of the topography and the seasonality of weather events and lifecycles have contributed to distinct microsite ecotopes.

The Roman empire profoundly affected communication between different peoples and over many centuries the circulation of goods, seeds and animals was encouraged. This probably played a relevant role at a local scale, compared to the input of new species after the discovery of the American continent.

The high ecodiversity of the Mediterranean region has been enhanced by a fine-grained use of the resources in time and space. This has probably ensured a historical sustainability for the population, although catastrophic events such as drought, floods, famine, war and disease have cyclically perturbed the system.

The fragility of the Mediterranean landscape is connected to its human stewardship, but resilient mechanisms have been adopted by plants, animals and systems.

Logging, grazing and fire are the main sources of disturbance in the Mediterranean. Logging activity, mainly for charcoal and timber, has been moderate thanks to the use of concrete to build houses.

Grazing has been very heavy in most mountain ranges across the Mediterranean, and in many cases overgrazing has exposed the soil to erosive processes.

Fire has played a fundamental role in controlling open spaces in the Mediterranean. In fact, the maintenance of spaces has been a priority for the rural population at all periods, from the protohistoric period, when hunting and gathering activities were dominant, onwards.

The fire tolerance of most Mediterranean plants – well demonstrated by many authors – has probably created most of the actual plant associations (Naveh 1990).

The capacity of plants to react to seasonal stresses such as drought and fire is extraordinary across the Mediterranean. Bulbs, short growth periods and abundant seeding using abiotics such as wind and water and biotic vectors such as animal fur, are some survival strategies.

However, although these adaptations have been extensively studied by ethnobotanists and zoologists, the ecological aspects are largely unexplored. In fact, the complexity of abiotic, biotic and human interactions producing such tremendous ecological diversity appears relevant in the Mediterrean region. Such ecological diversity create a resilient and sustainable system.

The whole Mediterranean suffers from seasonal human overcrowding (mass tourism), especially along the coasts. The resulting coastal development has definitively destroyed most of the land–marine interfaces, and consequently an essential buffer zone in which energy and matter meet has been lost. The demand for food, water and energy has forced other regions, often mountainous ranges, to work as source areas for the coastal sink.

Some strategies have been refined during this long period of human–nature interaction and some examples will be presented from different parts of the Mediterranean basin from which large data sets are available.

From Portugal across Spain a belt of oak savanna called, respectively, *montado* and *dehesa*, has been created by humans. This landscape is extremely characteristic and unique, mimicking the African savannas (see Fig. 1.9). The *coltura mista* of the Apennines (Italy) (Vos and Stortelder 1992) (Fig. 6.4) is another example.

Additional examples can be described as the transition from mountain to lowland and vice versa.

The seasonality of the human use of Mediterranean resources is coupled with the seasonal phenology of many animals and plants.

Geophytes escape the human disturbance (grass cutting) by reducing their living biomass to underground bulbs, and many plants benefit from the transportation of seeds in sheep's wool.

Plants and animals have the capacity to react to every human disturbance activity, and secondary succession and/or mutual benefits are visible in every disturbance event.

Human disturbance over such a long period has produced definitive changes in the composition of plant and animal communities; nevertheless, a diffuse biodiversity still exists in overcrowded and overused sites across the Mediterranean. Some plants were probably induced into speciation by the urban habits of the older cities such as Athens and Rome (Celesti-Grapow 1995).

Animals and cultural landscapes have been particularly studied in recent years in the Apennine region (Italy) (Farina 1991, 1994, 1995). There is evidence that the open spaces of montane and submontane prairies and the terraced olive orchards and vineyards of *coltura mista* are preferred sites for many stopover trans-African migratory birds and north-central Europe wintering birds (Farina 1986a). Favourable microsite temperatures and an abundance of food resources are important factors attracting birds.

Most of the managed lowland and marshlands along the Mediterranean coast are key places for the breeding and/or wintering of many populations of wader birds. The Camargue and the Tuscan Maremma are the most famous sites, but from North Africa to Sardinia the coastal marshes attract many species of birds. The pastoral upland areas across the Mediterranean are important for the survival of wolves, bears and large scavenger birds.

The scenic value of these areas is directly connected to the high human-related ecological diversity, both ingredients for a human-quality landscape.

Around the world the cultural landscape has an additional value compared to recently developed or wild areas. In the future more and more nature will be transformed by human activity, and the challenge will consist in our ability to mimic natural processes and preserve fluxes of materials and energy.

A basic ingredient of a cultural landscape is the survival and healthy functioning of the natural processes, maintaining and restoring the fertility of soil after each harvest by adding livestock manure. The stream corridors and the edges should be maintained in a healthy condition, to create a barrier against wind erosion, thermal excursion and frost exposure. Edges are used contemporarily by many ecotonal species and more tolerant plants and animals.

Birds have been largely studied in cultural landscapes (Farina 1986b, Purroy and Rodero 1986) and special attention has been devoted to the capacity, especially of frugivorous species, to adapt to such landscapes. For instance, the blackcap (*Sylvia atricapilla*) and the song thrush (*Turdus philomelos*) are able to track fruit availability in the olive orchards of southern Spain. These two species move from harvested to unharvested stands, tracking and recognizing food abundance at different spatial and temporal scales.

Olive orchards are not homogeneous plantations: composition density and ripening vary according to microsite quality, cultivar type and harvest timing. Consequently this landscape is extremely patchy and birds can compensate fruit availability across the mosaic. This capacity probably evolved in natural conditions before the human intrusion of the last millennia, allowing many frugivorous birds to flourish in the Mediterranean cultural landscape. Birds have been found more abundant and communities more diverse in rural areas than in woodlands in a sub-Mediterranean landscape of northern Italy (Farina 1997). More about the value of this landscape will be presented and discussed in the next section.

6.8 PATTERNS AND PROCESSES IN LAND ABANDONMENT

6.8.1 Introduction

Fragmentation and land abandonment are two main landscape processes that over a short time have

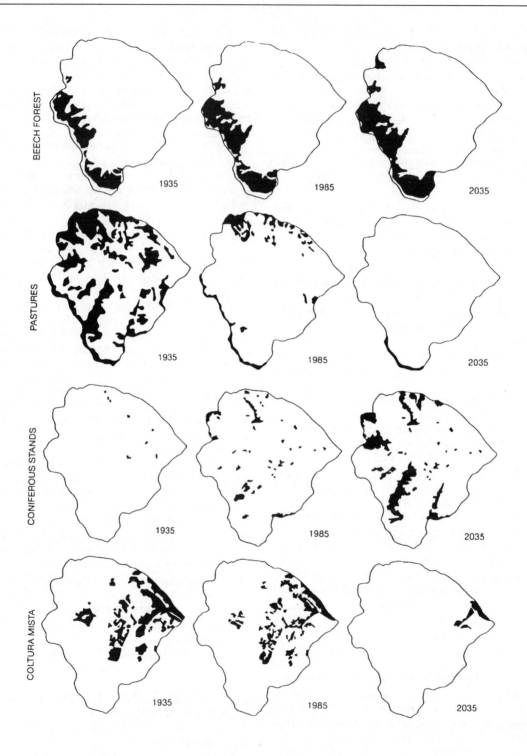

Figure 6.4 Effects of land abandonment in a Tuscan landscape (Solano basin) from 1939 to the 2035 prediction (from Vos and Stortelder 1992, with permission).

modified large parts of our planet. Fragmentation (see Chapter 4) is produced by the removal of pristine or secondary forests to gain new spaces for agriculture. Land abandonment is very common in industrialized countries (Baudry 1991, Baudry and Bunce 1991).

The attention devoted to fragmentation is far greater than that paid to land abandonment, but although there is a common sense that fragmentation causes a loss of biodiversity, the effect of land abandonment on biodiversity and on ecological diversity in general is not well understood.

In some regions this process may have beneficial effects on biodiversity, especially if the land is abandoned after intensive use, but in some regions such as the Mediterranean, land abandonment is causing not only a change in biodiversity but a real decrease in the number of plants and animals. In this landscape, which is generally considered a cultural landscape (*sensu* UNESCO; see Chapter 7 for more details) (von Droste *et al.* 1995), a fine-grained mosaic of fields and woodlot, a seasonal disturbance regime has for thousand of years maintained a great variety of plants and animals. In this case human stewardship has created a land mosaic extremely rich in local conditions.

Human societies and their occupied environment are closely connected, each social change coinciding with a comparable change in landscape structure and function. These modifications occur differently according to the history of each country or group of countries, the continent and the climatic conditions. It is hard to make generalizations on these processes. However, in an attempt to clarify these complex phenomena key cases in different continents will be described, with special emphasis on the Mediterranean region.

Before this can proceed, we must do away with dogmatic assumptions that terrestrial ecosystems have been influenced by human activity only in the last few centuries: such was also extensively the case in the remote history of our ancestors.

Human impact on the environment can be perceived at different spatial and temporal scales, and the landscape scale is the most efficient to describe this change.

Land abandonment is the most conspicuous process, coupled with development. Generally people move from the uplands and bad lands to lowlands and industrialized areas. The landscape dominated over many millennia by human use is abandoned and a secondary succession modifies vegetation cover and con-

sequently animal assemblages. This is so generalized and common an event in most industrialized countries that little attention has been devoted to it by ecologists.

The structure of the landscape and the effects on plants and animals are the most common issues. In particular across the Mediterranean land abandonment, especially on mountain ranges, has been very common and widespread over the past 50 years (Farina 1991, 1994) (Fig. 6.5).

Figure 6.5 Effects of 50 years of land abandonment in a northern Apennines ridge (Mt. Prado, 2000 m a.s.l.). (**a**) Image of 1954. The large amount of bare or thin grass layer and the topographic relief is easily appreciated. (**b**) The secondary succession occurring between 1954 and 1989 has created dense beech forests and the regrowth of shrubs (*Myrtillus*) above the tree line (about 1500 m). Clearing in the beech forest has been filled by a very rapid succession.

Along the northern Apennines the land abandonment has been of different pace and intensity according to elevation, topography and local human community histories. As in most mountainous catena in Europe, past deforestation had created a patchy landscape and land abandonment has encouraged the woods to recover (Fig. 6.6).

The uplands were important pastures for livestock, especially sheep. However, this pastoral activity has completely vanished and the grass cover faces a quick change in vegetation composition.

One relevant consequence of land abandonment is the change in landscape mosaic owing to woodland recovery. Generally this process is not linear and in many cases is interrupted by new disturbance regimes, such as fire. In this case, although the vegetation is adapted to fire – but not to too frequent fires – when a fire event is repeated within a short time in the same place, the secondary succession produces a complex cover of weeds, parasitic and epiphitic plants. These conditions are less favourable to maintaining a high level of biodiversity.

Fires of human origin are probably a consequence of the loss of the economic value of the land. This hypothesis has recently been confirmed by observations of fire incidence in the northern Apennines. Owing to the oil crisis in the last 80 years wood has been discovered as a cheaper resource for house heating. Coppice practices abandoned more than 40 years ago were restarted, causing an immediate decrease in fire occurrences. This patterns has two distinct components, one of which is linked to a changed human attitude toward the

woodland: the discovery of an economic value has created more concern regarding fire. Secondly, the coppicing has created conditions less favourable to fire diffusion because of the presence of more forest roads and young green stands.

In many areas of Spain the traditional silvopastoral activity based on a shifting agricultural system combined with grazing is quickly vanishing, especially in areas with less favourable soil. In the past a long fallow period and the transformation of woodland into park-like savanna (dehesa) have been good strategies to reduce fire risk and ensure a minimum biomass for livestock and a modest yield from crops. The dehesa assures biomass for livestock through the oak leaves and fruit, and from the legume-rich grass layer. Dehesa has a tree density of 40–100 per hectare. The abandonment of the dehesa is producing a *matorralizacion*, with the development of coarse grasses and shrubs more prone to fire risk. This decreases biodiversity as in other Mediterranean areas (Naveh 1974, Joffre *et al.* 1988, Gonzales Bernaldez 1991).

6.8.2 The effect of land abandonment on fauna

The recovery of woodland and the loss of open spaces determine profound changes in animal assemblages in terms of abundance and species. In the Mediterranean most of the relevant fauna live in open spaces; their reduction produces local extinction, rarefaction and fragmentation of populations. The same process is observed with forest fragmentation.

The effects on the land mosaic (Vos and Stortelder 1992), birds and mammals (Farina 1994, 1995, 1997) are well documented. In some transition phases land abandonment creates a greater landscape heterogeneity, but the succession moves so fast that in less than a decade grass layers are transformed into dense shrublands, unattractive to most vertebrates and arthropods such as butterflies.

Whereas most open-space birds (from partridges to stonechat) are being dramatically reduced by land abandonment large mammals are reclaiming areas previously hostile and from which human

Figure 6.6 Effect of rural abandonment. Secondary succession develops within a short time forbs and shrubs. In this image terraced upland fields are invaded by ferns and *Juniperus communis*, and are moderately grazed.

Figure 6.7 Distribution of roe deer (*Capreolus capreolus*) (**a**) and wild boar (*Sus scrofa*) (**b**) in 1911 (left) and in 1987 (right) in Italy. The diffusion of these mammals is directly connected with the abandonment of hilly and mountain ranges (from Chigi (1911) and Perco (1987), quoted by Apollonio 1996).

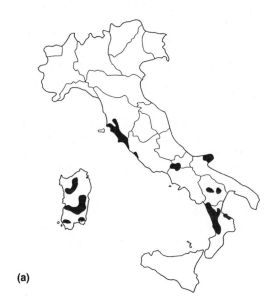
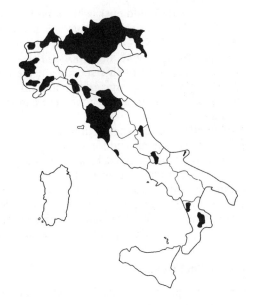

(a)

(b)

competition had driven them away (Apollonio 1996) (Fig. 6.7).

The spread of wild boar is becoming common in all of southern Europe, and this species, thanks to its adaptability and fertility, has become a true pest in many regions, with little possibility of human control. The effect of this species on the environment is very great: large herds of wild boar can 'plough' a prairie in a few hours and can seriously damage crops.

The effect of digging on upland prairies and the effects over medium and long periods on plant and animal communities is a debatable question for ecologists. In fact, the change in disturbance has been dramatic: only decades earlier they were heavily grazed by goats and sheep; for a few decades they were completely free from any disturbance, and now again they are strongly disturbed by wild boar.

6.8.3 Conclusions

The consequences of land abandonment on the landscape are:

- Disappearance of or reduction in cultural diversity;
- Reduction or extinction of many species of plants and animals;
- Change of the mosaic grain from fine to coarse;
- Reduction of open spaces and their transformation into dense shrublands;
- Increase of connectedness of forested patches;
- Decrease or disappearance of ecotonal synanthropic species;
- Decrease of the human-perceived scenic value of the land mosaic.

6.9 SUMMARY

- Changes in the landscape as a general rule depend mostly on disturbance, rate of recovery after disturbance, size and extent of disturbance and size of the landscape.
- Stability in landscapes is a matter of scale. Large-scale landscapes show small changes but at a small scale the changes may occur rapidly.
- Self-organizing mechanisms can maintain the shape of a landscape for a long time, but this largely depends on the geographical context.
- A hierarchy of shaping factors have an influence on landscape mosaic, from climate to soil and animal interactions.

- In human-dominated landscapes many factors change the landscape structure and functioning (e.g. agricultural intensification, land abandonment, fire suppression, deforestation, grazing, development).
- The Mediterranean landscape, owing to its history of nature–human interactions, is particularly vulnerable to dramatic changes in socioeconomic processes.
- Between the multitude of factors negatively influencing the quality of the Mediterranean landscape (e.g. industrialization, mass tourism, coastal erosion), land abandonment appears a point of concern.
- Land abandonment, especially in mountain ranges, is transforming the landscape grain from fine to coarse, from diverse and heterogeneous to homogeneous and monotonous, and affecting the composition and abundance of plant and animal communities.

SUGGESTED READING

Trabaud, L. (ed.) *The role of fire in ecological systems.* SPB Academic Publishing, The Hague, 1987.

Goldammer, J.G. and Jenkins, M.J. (eds.) *Fire in ecosystem dynamics. Mediterranean and Northern Perspectives.* SPB Academic Publishing, The Hague, 1996.

Kalin Arroyo, M.T., Zedler, P.H., Fox, M.D. (eds.) *Ecology and biogeography of Mediterranean ecosystems in Chile, California, and Australia.* Springer-Verlag, New York, 1995.

REFERENCES

Apollonio, M. (1996) Evoluzione dell'ambiente e delle attività antropiche nelle aree appenniniche in relazione alla presenza del lupo (*Canis lupus*). In: Cecere, F. (ed.) *Atti del Convegno dalla parte del lupo.* Atti and Studi del WWF Italia, n. **10**: 1–160.

Bailey, R.G. (1995) *Description of the ecoregions of the United States.* USDA, Forest Service. Miscellaneous Publication No 1391.

Baker, W. L. (1992) Effects of settlement and fire suppression on landscape structure. *Ecology* **73**: 1879–1887.

Baudry, J. (1991) Ecological consequences of grazing extensification and land abandonment: role of interactions between environment, society and techniques. *Option Mediterranéennes, Série Seminaires* 15: 13–19.

Baudry, J. and Bunce, B. (eds.) (1991) Land abandonment and its role in conservation. *Options Mediterranéennes* 15

Belsky, A.J. (1995) Spatial and temporal landscape patterns in arid and semi-arid African savannas. In: L. Hansson, L. Fahrig, G. Merriam (eds.), *Mosaic landscapes and ecological processes*. Chapman & Hall, London, pp.31–80.

Caravello, G.U. and Giacomin, F. (1993) Landscape ecology aspects in a territory centuriated in Roman times. *Landscape and Urban Planning* 24: 77–85.

Celesti Grapow, L. (1995) *Atlante della flora di Roma*. Argos Edizioni, Roma

Farina, A. (ed.) (1986a) *First Conference on birds wintering in the Mediterranean region*. Supplemento Ricerche Biologia della Selvaggina, X, Bologna.

Farina, A. (1986b) Bird communities wintering in northern Italian farmlands. *Supplemento Ricerche Biologia della Selvaggina*, X:123–135.

Farina, A. (1991) Recent changes of the mosaic patterns in a mountainous landscape (north Italy) and consequences on vertebrate fauna. *Options Mediterranéennes* 15: 121–134.

Farina, A. (1994) Effect of recent changes of the summit landscape on vertebrate fauna of northern Apennines. *Fitosociologia* 26: 171–175.

Farina, A. (1995) El abandono de tierras en un paisaje mediterraneo y las nuevas perspectivas de gestion holistica. In: Pastor-Lopez, A. and Seva-Roma, E. (eds.), *Restauracion de la cubierta vegetal en ecosistemas mediterraneos*. Instituto de Cultura Juan Gil-Albert (Diputación Provincial de Alicante), pp. 193–212.

Farina, A. (1997) Landscape structure and breeding bird distribution in a sub-Mediterranean region facing land abandonment. *Landscape Ecology* (in press)

Franklin, J.F. and Forman, R.T.T. (1987) Creating landscape patterns by forest cutting: ecological consequenées and principles. *Landscape Ecology* 1: 5–18.

Gonzales Bernaldez, F. (1991) Ecological consequences of the abandonment of traditional land use systems in central Spain. *Options Mediterranéennes* 15: 23–29.

Haralick, R.M., Shanmugam, K., Dinstein, I. (1973) Textural features for image classification. *IEEE Transactions on Systems, Man, and Cybernetics* SMC-3: 610–621.

Ishe, M. and Norderhaug, A. (1996) Biological values of the nordic cultural landscape seen in a comprehensive perspective. In: Ishe, M. (ed.) *Landscape analysis in the nordic countries*. Swedish Council for Planning and Coordination of Research, Stockholm, Sweden, pp. 30–53.

Iverson, L.R. (1988) Land-use changes in Illinois, USA: The influence of landscape attributes on current and historic land use. *Landscape Ecology* 2: 45–61.

Joffre, R., Vacher, J., De Los Llanos, C., Long, G. (1988) The dehesa: an agrosilvopastoral system of the Mediterranean region with special reference to the Sierra Morena area of Spain. *Agroforestry Systems* 6: 71–96.

Kratz, T.K., Benson, B.J., Blood, E.R., Cunningham, G.L., Dahlgren, R.A. (1991) The influence of landscape position on temporal variability in four North American ecosystems. *American Naturalist* 138: 355–378.

Krummel, J.R., Gardner, R.H., Sugihara, G., O'Neill, R.V., Coleman, P.R. (1987) Landscape patterns in a disturbed environment. *Oikos* 48: 321–324.

Milne, B.T. (1991) Lessons from applying fractal models to landscape patterns. In: Turner, M.G. and Gardner, R.H. (eds.), *Quantitative methods in landscape ecology*. Springer-Verlag, New York, pp.199–235.

Naveh, Z. (1974) The ecological management of non-arable Mediterranean uplands. *Journal of Environmental Management* 2: 351–371.

Naveh, Z. (1990) Fire in Mediterranean – A landscape ecological perspective. In: Goldammer, J.G. and Jenkins, M.J. (eds.) *Fire in ecosystem dynamics*. SPB Academic Publishing, The Hague, pp. 1–20.

Naveh, Z. and Vernet, J.L. (1991) The paleohistory of the Mediterranean biota. In: Groves, R.H. and Di Castri, F. (eds.) *Biogeography of Mediterranean Invasions*. Cambridge University Press, Cambridge, pp. 19–32.

Purroy, F.J. and Rodero, M. (1986) Wintering of wood pigeons (*Columba palumbus*) in the Iberian peninsula. *Supplemento Ricerche Biologia della Selvaggina*, X: 275–283.

Simpson, J.W., Boerner, R.E.J., DeMers, M.N., Berns, L.A. (1994) Forty-eight years of landscape change on two contiguous Ohio land-scapes. *Landscape Ecology* **9**: 261–270.

Skinner, C.N. (1995) Change in spatial characteristics of forest openings in the Klamath Mountains of northwestern California. *Landscape Ecology* **10**: 219–228.

Turner, M.G. and Ruscher, C.L. (1988) Changes in landscape patterns in Georgia, USA. *Landscape Ecology* **1**: 241–251.

Turner, M.G., Romme, W.H., Gardner, R.H., O'Neill, R.V., Kratz, T.K. (1993) A revised concept of landscape equilibrium: disturbance and stability of scaled landscapes. *Landscape Ecology* **8**: 213–227.

von Droste, B., Plachter, H., Rossler, M. (eds.) (1995) *Cultural landscapes of universal value*. Gustav Fischer, Jena.

Vos, W. and Stortelder, A. (1992) *Vanishing Tuscan landscapes. Landscape ecology of a sub-Mediterranean-montane area (Solano Basin, Tuscany, Italy)*. Pudoc Scientific Publishers, Wageningen.

White, M.A. and Mladenoff, D.J. (1994) Old-growth forest landscape transitions from pre-European settlement to present. *Landscape Ecology* **9**: 191–205.

Wiens, J.A. (1995) Landscape mosaic and ecological theories. In: Hansson, L., Fahrig, L., Merriam, G. (eds.), *Mosaic landscapes and ecological processes*. Chapman & Hall, London.

Principles of landscape conservation, management and design

7.1 INTRODUCTION

Seminatural, rural and agricultural landscapes, interdispersed urban centres and relevant infrastructures such as highways, bridges, dykes, electricity poles etc. are widely distributed in the world. The emergent characteristics of these landscapes are in synthesis with the fragmentation of natural remnant vegetation, the high heterogeneity, the interdispersion of different matrices of fields. Size, shape and the spatial arrangement of the patches remain relevant for ecological processes. The overlap of technological infrastructures (roads, bridges, railways) with natural structures such as rivers, lakes, valley bottoms and ridges creates hindrances to many ecological processes, such as soil erosion and deposition, water flux, animal movements and plant dispersion.

This chapter deals with the importance of landscape ecology as the scientific basis for the study, planning and management of seminatural, rural and agricultural landscapes. The capacity of landscape ecology to track ecological processes across a range of spatial, temporal and cultural scales allows us to understand the real or potential effects of human land use and planning. An explicit application of landscape ecology principles to planning should permit us to mimic and/or preserve natural processes. It is important in conservation strategies in pris-

tine environments as much as in human-influenced landscapes. So, in this chapter our goal is to clarify the role of landscape ecology in environmental issues, which, owing to the increasing deterioration of the biosphere, is of primary concern to ecologists but also to politicians and decision makers.

Landscape ecology can improve the anthropocentric approach to these issues in both undisturbed and disturbed landscapes. The principles of landscape ecology can be used to plan or manage key species, forest remnants, the network of edges and woodlots in croplands, or to influence and direct urban development.

Special attention has been devoted to the value of cultural landscapes, focusing on their importance in conserving biological diversity and the diversity of many ecological processes, and recognizing the inherent ecological value of some man-shaped landscapes.

The importance of the cultural landscape forces us to place this issue first because most of the strategies that can be used to conserve pristine or deteriorating landscapes can find an immediate application in the cultural landscape. Owing to the broad spectrum of possibilities, only exemplary key studies will be presented in an attempt to cover the most significant issues in nature conservation, management and planning.

7.2 LANDSCAPE EVALUATION

A focal point in landscape management is assessment of the value of the landscape and finding criteria by which to evaluate its components. Naturalness has been proposed by Anderson (1991) as a conceptual framework that is synonymous with intactness or the integrity of an ecosystem. This author proposed three indices of naturalness:

1. The degree to which a system would change if human interference were removed;
2. The amount of cultural energy required to maintain the functioning of the system as it currently exists;
3. The complement of native species that remain in an area compared with those present prior to settlement.

The concept of naturalness is popular because the word is attractive, but it appears difficult to apply because often the exclusion of humans from nature depresses biological and ecological diversity. The general assumption that human influence on the landscape is always negative should be contradicted: in many cases human influence does indeed decrease the value of a site, but in many cases it is necessary to maintain biodiversity (Gotzmark 1992).

Most of the long-term human-modified landscapes can be considered cultural landscapes. In these human stewardship has created feedback with many ecological processes, and the recreation of a natural landscape appears more of a dream than a scientific reality. From at least the Holocene period cultural landscapes have extended over large parts of Europe and elsewhere.

7.3 THE CULTURAL LANDSCAPE

7.3.1 Definition

There are many definitions of a cultural landscape. We consider as pertaining to this type a landscape that has been changed in some parts by a long-term human disturbance regime by which a unique assemblage of patterns, species and processes has been created. So, a cultural landscape is a human-dominated landscape in which the patch arrangement and their quality and function have been the result of millennia feedback between natural forces and humanity. A cultural landscape reflects the interactions between people and their natural environment, and is a complex phenomenon with both a tangible and an intangible identity (Plachter and Rossler 1995).

In 1991 the Unesco Secretariat proposed guidelines to identify a valuable and endangered cultural landscape: '...be an outstanding example of a cultural landscape resulting from associations of cultural and natural elements significant from the historical, aesthetic, ethnological or anthropological points of view and evidencing a harmonious balance between nature and human activity over a very long period of time which is rare and vulnerable under the impact of irreversible change' (reported in von Droste *et al.* 1995).

Generally these landscapes have a complicated structure represented by a fine-grained mosaic in which physiotopes have been well localized and utilized in different ways by agriculture, forestry and pastoralism. Often slopes are transformed into terracettes, which reduce soil erosion and facilitate the practice of agriculture by improving nutrient retention (Fig. 7.1).

Cultural landscapes are a good model for testing the possibility of expanding humanity in a natural environment without dramatic resource depletion and irreversible habitat perturbation. This could be seen as a Utopian ideal, but the lessons that cultural landscapes teach should not be ignored. There is an urgent need – not an option – to find a balance between healthy human development and sustaining the ecosphere (Halladay and Gilmour 1995).

7.3.2 Interactions between natural and cultural landscapes

A cultural landscape is the product of change created in natural landscapes by long-term human influence. A cultural landscape requires the human stewardship to be maintained, and for this reason is fragile and returns to its natural shape when human interference disappears or is reduced.

When land abandonment occurs in a cultural landscape, the structure of that landscape is transformed. In the Mediterranean upland landscape, terracettes are progressively broken up and fertile soil is lost by water erosion. Livestock, which is generally used to maintain some openness in the landscape after agriculture abandonment, causes terrace wall degradation by trampling (e.g. in the Mediterranean deer and wild boar are particular culprits).

Figure 7.1 The terraced riverine landscape near Moncigoli, northern Apennines. Example of a cultural landscape: the Apennines *coltura mista* characterized by small fields separated by vineyards and surrounded by dense woodlands.

Relevant differences can be found when natural and cultural landscapes are compared. The structure of a cultural landscape is often more patchy than a natural landscape, or more homogeneous. In a cultural landscape the intermediate level is often lacking, but it is not possible to generalize for all cases.

Cultural landscapes have more linear structures such as hedgerows, or more open spaces than the original landscape. However, in some cases the opposite applies. A desert oasis has more plants than a natural desert spring; mountain farmland has fewer trees than a mountain forest but more trees than a natural mountain prairie.

There is an infinite number of types of cultural landscape around the world, but all are strictly structured according to local tradition, use and resource sustainability. However, this sustainability is time dependent according to the regional history.

Cultural landscapes are generally created by sedentary populations, but there is no reason to exclude landscapes modified by nomadic populations, as in the Mongolian steppes, for example. Here the sustainability is assured by a shifting grazing mosaic of livestock and the stewardship is manifested not by digging the soil or pruning trees, but by maintaining light livestock grazing, thereby reducing trampling and biomass consumption.

Some regions of the earth, like the Mediterranean, can be considered a cultural landscape. Here the long-term interaction of populations and environment has produced irreversible changes in biological as well as ecological diversity.

In many regions land use change and/or land abandonment has produced heavy modifications in the cultural landscape and the disruption and disappearance of these valuable landscapes causes concern in the authorities.

The birth of countryside heritage centres is a timid reply to a diffuse problem of land management which in this technologically oriented age seems hard to solve.

Often the cultural landscape has more biological species than a natural landscape, and in other cases the degradation produces a poor-quality landscape in which there is species reduction (Farina 1995).

We must be careful when we consider the natural value of the landscape. In regions like the Mediterranean biodiversity was depleted thousands of years ago, and there are no new species to colonize empty land niches.

7.3.3 Fragility of the cultural landscape

We have observed that cultural landscapes are fragile and need human stewardship to be maintained. During the land abandonment process, especially in dry regions, the gradual evolution from an anthropogenic patterning to a more natural mosaic is disturbed by fire.

Fires are very common in the Mediterranean basin, and in dry regions in general. Although most Mediterranean plants are fire tolerant and many actually require fire to complete their lifecycle, in most of the Mediterranean basin the increasing frequency of human-induced fires dramatically reduces the ability of the system to incorporate the disturbance. In many cases large ranges are deforested and soil is exposed to erosion during heavy rains.

Apparently the presence of roads encourages such types of fires, and also land abandonment. How can fires be controlled? In the Mediterranean this can be done by evaluating the economic value of woodlands. This seems a very simplistic remedy, but in many cases is the reality. Along the northern Apennines there is much evidence that the frequency of fires has decreased over recent decades when, because of the oil crisis, logging activity has been resumed and the woodlands discovered to be an important source of income. Logging activity in this region consists in clearing young stands (25–50 years old) of variable size, generally to a maximum of 20–30 ha. In this way only a limited portion of soil is exposed at any one time, and the following year a luxuriant secondary succession reclaims the understorey, preventing water flashing. In this way dense homogeneous woodlands are transformed

into a checkerboard of woodlots of different ages, structure and resource availability.

The availability of cadastral maps allows us to predict the heterogeneity of woodland in many parts of the northern Apennines, creating a wonderful tool, especially if the historical information from old cadastral maps is compared with the actual using GIS facilities (Fontanelli, pers. com.) (Fig. 7.2).

7.4 PRINCIPLES OF LANDSCAPE MANAGEMENT

7.4.1 Introduction

One of the main goals of landscape ecology is to study the structure of the spatial mosaic and its effects on the ecological process. Organisms, energy and resources are distributed patchily in the environment, and this distribution is important for most ecological patterns and processes. Complex mosaics are crossed by organisms, energy, nutrients, water and disturbance processes, and all these elements are influenced by landscape heterogeneity.

Landscape ecology studies the complexity created by the unequal distribution of energy, resources, organisms, predatory risk, populations and communities. Even a bottle of pure water is not homogeneous, since if it were possible to track every molecule, each would have a specific energy and be influenced by neighbouring molecules.

As argued by Noss (1995), the management of species site by site or ecosystems species by species is not a promising approach. Landscape ecology considers sites not in isolation from each other, and recognizes that scaled investigations are definitely the most intriguing and powerful approach to understanding complexity (Haber 1990). In fact, homogeneity and heterogeneity are two different ways of seeing the environment. Homogeneity often refers to the quality of adjacent patches, as when distinguishing prairie from forest using a satellite image classification, but if we look just at the prairie level, using aerial photographs for example, this environment appears heterogeneous.

Landscape scale is one of the most efficient approaches to ecosystem management. The landscape embraces a large piece of land where most of the natural and socioeconomic processes occur, and it can be considered to contain most of the patterns and processes of interest to us.

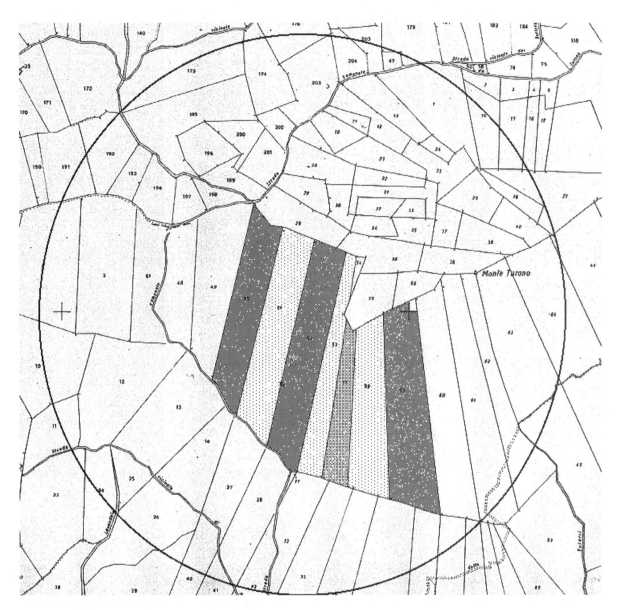

Figure 7.2 Example of woodland property mosaic depicted by cadastral maps. This mosaic is maintained by periodic logging and the combination of differently aged stands largely depends on the history of properties. In this case the socioeconomic structure of a local society has a strong influence on the patterning of woodlands and consequently on many ecological processes.

The landscape scale considers the watershed or other units, such as ecotope, micro-, meso-, macro- and megachore, as the fundamental area on which direct management actions are determined. Often the landscape hierarchy is overlapped by social and administrative boundaries: e.g. in Italy and in many other countries in Europe there is a good correspon-dence between landscape hierarchy and social and administrative entities: ecotope/field/cultivation, microchore/farms, mesochore/parish/communes, macrochore/provinces/regions.

In only a few years land managers and conserva-tionists have shifted from an autoecological approach to a synecological one, and finally to a landscape management approach. In some literature this last is called the ecosystem approach, but it is clear that the authors were dealing with the landscape.

The landscape approach assumes such importance in land management because human action in conservation, restoration and general management acts at a landscape scale.

7.4.2 The importance of watershed scale management

Watershed scale seems a very appropriate approach to efficient land management. A watershed may be considered a multifunctional unit in the landscape, and unique processes like water and mineral fluxes are distinct.

The case of Fymbos in the region of Cape Town encapsulates very well the concept of watershed management. Recently, Fymbos' unique vegetation and spectacular flora have been threatened by alien plants that produce an increase in biomass. This causes a decrease in runoff from the catchment area (van Wilgen *et al.* 1996). The invasion of European, North American and Australian weeds has created a complicated situation in which not even fire can control the invaders. The fire regime that occurs regularly in this area finds a much greater fuel source, with more severe consequences for plants. These fires cause landscape degradation by increasing soil erosion, which is also increased by the loss of indigenous fire-tolerant plants that cannot protect the soil by rapid recovery.

One consequence of this landscape change is the reduction of water availability for the lowlands. In this case, managing the environment at watershed scale should reduce the invasion of alien species, reduce fire severity and reduce water consumption by plants, thereby making more water available for the environment and people in the lowlands.

7.4.3 The role of keystone species in landscape management

The term keystone species was used for the first time by Paine (1966, 1969), studying the rocky intertidal zone. A keystone species plays a fundamental role in shaping and structuring a community (Mills *et al.* 1993). From this original definition the concept has been enlarged to include processes (Table 7.1).

The role of keystone species in the landscape may be of great importance. For example, in the Great Plains buffalo have modified the landscape by creating open spaces and trails, and reduced the grassland biomass. The same occurs in the African savanna, where the manure, urine, defoliation and trampling of huge numbers of large herbivores maintain the open structure.

When a keystone species disappears because of natural or human causes many species collapse and the landscape changes. In management policies keystone species must also be considered for their relatively inexpensive use.

7.5 NATURE CONSERVATION AND LANDSCAPE ECOLOGY

7.5.1 Introduction

Nature conservation generally focuses on specific areas of land and distinguishes between two main types of protected area: the national or international parks, and reserves.

National/international parks are heterogeneous areas in which generally different biotopes are present and in which some processes are sufficiently conserved. Reserves are generally patchy and smaller, being surrounded by an agricultural or developed matrix.

Nature conservation plans are generally achieved following three main approaches:

- Conservation of threatened plants and animal populations;
- Conservation of representative biotopes (a piece of land with a particular type of nature), including communities and ecological processes;
- Conservation of areas with high biological and/ or ecological diversity.

Generally conservation policies tend to conserve structures rather than processes.

Landscape ecology can be useful for nature conservation because it takes into consideration the spatial arrangement of habitats, and consequently considers structures and processes as perceived by the different species. This perception often does not coincide with human perception, creating difficulties when management action is required. A compromise should be found, since in the future wild, remote and large areas will be increasingly rare because of human intrusion.

Table 7.1 Categories of keystone entities and the probable effects of their removal. (from Mills *et al.* 1993, with permission).

Keystone category	Effect of removal
Predator	Increase in one of several predators/consumers/competitors, which subsequently extirpates several prey/competitor species
Prey	Other species more sensitive to predation may become extinct; predator populations may collapse
Plant	Extirpation of dependent animals, potentially including pollinators and seed dispersers
Link	Failure of reproduction and recruitment in certain plants, with potential subsequent losses
Modifier	Loss of structures/materials that affect habitat type and energy flow; disappearance of species dependent on particular successional habitats and resources

Whereas the use of indicator species appears difficult, the selection of guilds as management units seems more appropriate (see Wilcove 1989).

The loss of natural areas is a diffuse trend around the planet and the preservation of valuable habitats or biotopes is achieved by preserving small remnants of such landscapes. Some reserves represent an archipelago of surviving habitats, but are often islands in a matrix of agricultural or developed areas. From this isolation can arise the risk of stochastic extinction. Species generally need different types of habitat during their lifecycle, and often this is not sufficiently provided for in reserves; also, reserves are bounded by an artificial, human-transformed landscape.

The contrast between reserves and the surrounding matrix could create climatic stresses, especially at the edges, a concentration of predators and a greater influence of alloctone species on indigenous ones. Intensive agriculture in the matrix will offer more resources to the reserve than extensive agriculture and forestry. A change in keystone species – the bottom keystone species (plants) as well as top keystone species (predators), the first for climatic change and the second for landscape alteration – can modify the overall community.

Several examples are given by Hansson and Angelstam (1991) of the consequences of fragmentation and isolation of forest areas in a crop landscape. Predation increases; the influence of seed consumers such as mice can reduce the regeneration capacity of the forest. Some keystone species like the woodpecker, which bores holes in trees and allows other hole-nesting species to breed, in small forests plots can encourage the settlement of jackdaws (*Corvus monedula*), which are wholly dependent on the agricultural landscape. Small remnants of flower-rich meadows have low numbers of pollinators (hymenopterans and lepidopterans), and this produces a cascade effect on the plant community.

For instance, the number of herbivore insects on blueberry (*Vaccinium myrtillus*) is higher in wet, interior forest patches. This is probably due to the greater palatability of this plant compared to those in lighter, drier forests (low content of carbon-based phenolics, which are antiherbivore substances). The abundance of insects favours the presence of insectivorous birds such as flycatchers and grouse. The destruction of wet forests has consequences for many guilds and animal assemblages.

7.5.2 Landscape principles for natural reserves

Some basic principles for creating and maintaining natural reserves are unanimously agreed:

- Species richness increases with forest area. This is especially evident in tropical areas subjected to forest clearing for agricultural exploitation. Also, in temperate biomes forest remnants have more capacity to conserve biodiversity at reasonable levels.
- A continuous area has more native interior species than two or more small ones. This principle is especially true in North American temperate and boreal forests.
- In a forested area separate patches close to each other support more species than patches that are further apart. This principle is in line with the concept of connectivity.
- Disjunct patches connected by strips of protected area are preferable to fully isolated patches.
- Other things being equal, a circular reserve is better than an elongated one because the portion of interior habitat is larger.

These principles in turn recognize the importance of area, patch shape, connectedness and edge development attributes of the land mosaic as studied and modelled by landscape ecology.

7.5.3 Disturbance regime and reserve design indications

The main concern in nature conservation is the perpetuation of species, populations, communities and processes. Recently new approaches have focused on the perpetuation of landscape form and process.

Conservation planning must consider all levels of biological organization, from population to landscape. Landscape structure appears more and more important for nature conservation. In particular, large disturbance regimes seem fundamental to maintaining ecological processes (Baker 1992).

Disturbance occurs in all biomes, and its role in maintaining structure at the species, ecosystem and landscape scale is increasingly being recognized. The effect of disturbance on the landscape mosaic is discussed extensively in Chapter 3.

Disturbance regimes find a growing importance in reserve design and management. Baker (1992) focuses on some general principles by which disturbance represents an important component to the maintenance and perpetuation of target patches and landscapes.

To preserve species and processes it is necessary to have enough space in which natural disturbance can interact with ecosystems; the space is necessary to ensure the shifting of regeneration patches and the movement of species from different-quality patches. According to the source–sink model the strategy should be to localize the source area for a species and manage it with special attention. The preservation of climax communities can be assured by buffering them with successional stages. The biodiversity will be high, although this represents a mimic of natural conditions lost by human clearing and cultivation.

It is clear that the best conservation strategy is to preserve large natural areas, but often this is idealistic and we are faced with isolated remnant patches. In this case the conservation also of small areas may be important, but studies on the ecology and species-specific habitat requirements are necessary. In this sense landscape ecology, with its capacity to focus on the entire system, composed of crops, developed and natural mosaic, is of great use in taking into account more variables than the local perspective of nature conservation.

One prerequisite in designing a reserve is that it should contain a minimum dynamic area, that is, the smallest area in which under a natural disturbance regime internal recolonization sources are maintained. This means that the area should have a family of disturbance patches with a temporally stable structure. This means the incorporation of disturbance in the reserve bounds. In other words, the spatial arrangement of disturbed patches changes but the overall structure of the landscape is maintained, creating a shifting-mosaic steady state.

If the goal of management is the maintenance of the disturbance regime it is important to have an area enough large to incorporate the greatest disturbance. For this reason the choice of clusters of small reserves seems less favourable to the maintenance of internal disturbance dynamics.

In order to design a reserve composed of core, buffer and transition areas (see Harris 1984), it is important to avoid such reserves being islands in an ocean of intensive human use. Some contradictions arise when such a model is compared with two different goals, those of maintaining high diversity in the reserve and ensuring the disturbance regime. The buffer for maintaining the disturbance regime should be, for instance, a secondary succession close to old-growth forest in the reserve, but in terms of species conservation this buffer area could increase the number of alien species which are both habitat tolerant and predators. Equally, the fire risk is higher if a buffer of shrubland encircles a core area. In reality the role of the buffer should be both to mitigate undesired disturbance and to act as an area in which to manage disturbance.

Geographically it is important to place reserves where the disturbance begins, and also to include the areas in which it naturally dies out. The first requirement is necessary for managing disturbance and the second to maintain this regime politically. However, if the reserve has continuity with urban areas or managed forests, it is clear that a fire regime will never be adopted. For example, to maintain flood disturbance it is necessary to include in the reserve the upstream watershed where the disturbance begins, and also the downstream basin where the disturbance (in this case flood) is exported. A boundary can be placed between a windward and a leeward slope for wind disturbance. For fire disturbance bare soil, a lake or a stream seem good boundaries. Where there are no natural borders human breaks can be provided for this purpose.

So, the decision to locate forest reserves along riparian strips means that a fire regime is avoided. In this case it appears clear that conservation is focusing on species along a restricted temporal perspective, transforming the system into a habitat system. Strategies are also required to develop an ecosystem-level or landscape-level reserve system.

These principles can be exported to management outside reserves to maintain ecological diversity and the landscape in good health (Halladay and Gilmour 1995). For example, in the Mediterranean logging, grazing and fires have shaped the cultural landscape. In order to preserve these landscapes in places that are considered valuable it is important to ensure that these regimes are still working.

Recent land abandonment has changed these regimes and transformed a patchy landscape which was both valuable and suitable for many species into a more homogeneous system in which the disturbance regime has been modified. For instance, fire produces serious modifications in the system owing to the high connectivity of woodlands, connectivity that in the past was less owing to crop patches. The green biomass was consumed by livestock and the fire risk smaller than at present.

Baker (1992) points out three main strategies to handle disturbance focused on nature conservation: surrogation, suppression and prescription.

To surrogate a disturbance means to use a disturbance that mimics another: for instance, logging can mimic fire. But while size, shape, timing and spatial distribution can be surrogated, it is hard to surrogate the intensity and the final result can be far from expectations.

The suppression disturbance regime is a common practice in reserves. Fire, grazing and logging generally are suppressed. The suppression generally produces a coarse-grained landscape rather than the fine-grained landscape that pertains if disturbances are maintained. Again, the cultural landscape of the Mediterranean uplands is an example (Farina 1996). The human disturbance suppression has created a coarse landscape, substituting the *coltura mista*, a patchy, park-like mosaic, for homogeneous woodland stands. The suppression control changes the structure of the landscape. For example, a dam along a stream modifies the dynamics of the entire stream downwards, reducing disturbance by flood.

The disturbance prescription is becoming a popular practice, for instance in managing natural prairies. However, owing to the modest size of most of these prairies (especially in the USA), fire prescription appears an insufficient long-term strategy, as argued by Baker (1992). Often the disturbance is produced at the same season and with a rotation to cover the entire area. Generally this does not occur in nature, and for this reason the long-term effects are unpredictable. This regime can alter the distribution and the behaviour of species living in ephemeral patches

created by a stochastic disturbance regime. And finally, small disturbances are not surrogates for a large disturbance. But to maintain small reserves it is necessary to reduce the risk of large disturbances that could destroy the entire reserve.

Intermediate transitional disturbance regimes could be utilized to manage systems in which fire suppression has created a biomass accumulation.

7.5.4 Interrefuge corridor design

Often corridors are utilized as a framework to mitigate the effects of habitat fragmentation, although few data are available to design effective corridors (Harrison 1992).

In mammals natal dispersal is a sensitive period in which corridor availability assumes an important antipredator role. According to the predatory pressure there are species that move very fast from one group to another (e.g. deer), whereas other species, such as large carnivores, disperse until they find suitable unoccupied habitats. There is evidence that topographic features such as mountain passes, rivers and lakes may affect dispersal paths, but this information is not enough to predict the distance and direction of dispersal.

If a corridor has enough suitable habitats for a species it is reasonable that individuals can disperse easily along the corridor. Other factors can play a role in the effectiveness of a corridor: for instance, the seasonality of movement of males and females, the presence of human settlements, roads and other human creations, such as power lines, railways, aqueducts, pipelines etc.

Lindemayer and Nix (1993) studied the presence of arboreal marsupials in Australian corridors and found more individuals of larger species than small species in wildlife corridors. However, large marsupials are solitary and are food opportunists. Small marsupials live in groups with a colonial structure and consume widely dispersed food resources, so apparently narrow, linear-shaped corridors have strong effects on these species. A combination of habitat requirements must be taken into account when corridor availability is considered. Lastly, the context of the wildlife corridor in the landscape is important. The corridors that connect gullies with ridges have more species than vegetation belts confined to a single topographic position, such as a midslope. So, what appears important, at least for the animals studied by these authors, is an assessment of the site context, connectivity and social structure, diet, and the foraging patterns of target species.

Once again a single pattern or structure is not enough to explain the behaviour of vagile organisms, but it is necessary to take into consideration more structural and functional components at different temporal and spatial scales.

7.5.5 Hedgerow system to conserve biodiversity in rural landscape

Hedgerows have been a very common pattern in many rural landscapes around the world, but are being reduced by the increasing mechanization of agriculture before their role in maintaining diversity and stability has been fully understood.

Their structures are very different according to their function and the regional cultural context, ranging from wind protection and soil conservation in Europe to medical or religious purposes in Asia. Their primary role is to divide properties, to create enclosures for livestock, to provide domestic fuel, and finally to enhance the scenery.

Generally in traditional rural areas they compose a network that increases the connectedness between different locations.

Hedgerows play different roles in the landscape and are involved in many ecological processes acting at different scales. They can be considered, according to perspective, as true habitats, ecotones, corridors and buffer zones for soil nutrients and managed fertilizers (Burel 1996).

At the landscape scale hedgerows play a role in the control of water flux, nutrient leaching and wind flow, and act as a barrier to animal and plant dispersal.

In recognition of this important role hedgerow planting programmes are becoming popular, especially in western Europe after a dramatic simplification of the rural landscape in recent decades (Jorg 1994).

A recreational role is also accorded to many hedgerow systems, and they can mitigate deforestation in many countries. Finally, hedgerows can be seen as structures from which forest regeneration can rapidly develop after land abandonment.

7.6 CONSERVATION IN LANDSCAPE

7.6.1 Introduction

Conserving biodiversity or water quality or other natural resources in a landscape is a hard task for land managers owing to the complexity and dynamism met at this scale. The social cost of maintaining landscape-oriented conservation plans is a central point in the debate. To preserve large animals and ecosystems from extinction considerable economic sacrifices are necessary, as reported by Mann and Plummer (1993).

Conservation strategies can be activated using target species, populations and communities at patch, landscape and regional scales.

It is a common requirement for individual species to ensure the following:

- Size of habitat patches (grain size);
- Homogeneity of patches (grain evenness);
- Distribution of habitat patches (grain dispersion);
- Matrix surrounding patches;
- Connectivity between patches.

It must be considered that the grain size of patches is species specific. Different species have different sensitivities to the above-mentioned characters. Species such as deer, living in flocks, have a different perception of habitat complexity compared to solitary species. Habitat requirements can change according to season, physiological status, age and the quality of the occupied habitat.

Often the conservation of a target species can produce benefits to other species, especially if that species occupies a large portion of the landscape.

The conservation of target species requires a profound knowledge of autoecology, and often this is not available over the entire geographical range of a species. This remark is not intended to discourage the target-species approach, but to create awareness of the difficulties that can be found when moving in this direction, and that often more research is needed in advance.

Managing target species often means working in human-modified landscapes, with limitations and practical constraints such as property, the different regulations in the same region, and the overlap of management competencies.

These difficulties are common to other conservation approaches.

The conservation programmes in agricultural landscapes are increasing after the heavy impact of technology in recent decades. The aim is to maintain biodiversity and to ensure connectivity to fragmented subpopulations.

The restoration of field margins is currently very popular and positive results can be easily achieved, especially as regards arthropod diversity. This has economic implications because often, if a balanced invertebrate community is conserved, it may mean more biological control of pests and diseases. The conservation of vertebrates in rural landscapes can be achieved either by increasing the margins or by preserving the woodland remnants, ensuring good connectivity. The planting of mixed woodlots seems a very promising strategy to ensure a broad spectrum of resources over the year and for a more differentiated fauna. In should be clear that it is not possible to recreate natural patterns in farmlands, but the conservation of remnants, the increase in connectivity that may be a surrogate for a more natural matrix, may be a good strategy.

Conservation in disturbed landscapes in which the mosaic heterogeneity has been depressed by human activity may require a new disturbance regime. This approach seems very interesting, although not necessarily workable.

Prescribed fire is one of the most-used tools to ensure heterogeneity in remnant prairies in the USA, and to control biomass in coastal pine ranges. Although fire is a relatively inexpensive tool this practice can be difficult to handle when the managed areas are of limited size and the creation of an artificial checkerboard is not the plan.

It is not clear at present whether the habitat loss may be compensated for by careful planning of the landscape. Explicit spatial models can be very important to achieve this goal, but not all species have a population pattern that can be manipulated spatially. The spatially explicit population models (see also Chapter 8) are useful because they consider both species–habitat relationship and the spatial and temporal arrangement of habitat patches.

Generally these models are arranged for one or a few species, and adapting them to model biodiversity at the landscape scale represents a true challenge for landscape management (Turner *et al.* 1995).

7.6.2 Conservation of fragmented habitats and populations

Fragmented habitats and populations are very common in human-dominated landscapes and the survival of many species depends on the rate of connectivity between each fragmented habitat patch.

Habitat patches have to be considered heterogeneous in terms of area and quality, and the relationship between patches in a landscape, the spatial arrangement, the temporal change in landscape structure and the dispersal characteristic of the species are important components of this scenario.

Local populations can occupy a patch of the landscape and many local populations compose the natural population at regional scale. A habitat patch may be considered as a discrete area that a species uses for breeding or for obtaining other resources. Fluctuations in local abundance can cause local extinction, but at the regional scale if these local fluctuations are not synchronous the risk of regional extinction is negligible.

If extinction occurs in a habitat patch recolonization is an expected event. When we study this process at landscape scale it appears important to measure the landscape spatial structure, defined by Fahrig and Merriam (1994) as the spatial relationship between landscape components.

The characteristics of landscape spatial structures are:

- Size, shape and quality of patches. Larger patches have a higher persistence of a population. The dimension of a patch can influence the edge effect and the predator ratio. The size of a patch *per se* is not enough to characterize the degree of persistence of local populations, because two patches of the same size but with different edge amounts and shapes can support populations differently. Also, the patch quality is important. For example, the presence of old trees in a patch can increase the persistence of hole-nesting birds.
- The presence of corridors through the landscape. The function of these corridors is to maintain the connectivity between different local populations, and there are many evidences of the importance of connecting structures, such as fences, shrub strips and roads to encourage the movement of animals.
- The spatial configuration of landscape components is extremely important. In fact, the position of habitat patches in a hostile matrix seems more important than the dispersal routes. It appears important also to understand the rate of change of a landscape: if the landscape changes more rapidly than the population, the regional population will have difficulty in surviving.

It is clear that the value of small fragments of remnant habitats must be scaled with the overall importance of an area and is not an excuse to reduce large patches of undisturbed land. Nonetheless, it is important to recognize the role of small habitat fragments in preserving certain species, and its role is like a bank, to preserve genotypes rather than landscape dynamism.

Small areas cannot support large mammals but can play a relevant role for plants and invertebrates (Shafer 1995). Small reserves have little chance of preserving a species long term, and the biological diversity is lower than in the surrounding undisturbed areas. Small reserves created to protect rare butterflies, for instance, function for many decades without apparent loss of animals.

A fragment can be used as core area for a restoration project!

If a fragment is inadequate to preserve a particular species it could be enough for another species, and it is clear that small fragments may have a relevant role in a landscape. Managing small areas means that these fragments can be replicated in a landscape, protected by buffer zones and connected to each other by usable corridors.

Although forest or prairie fragments are not the optimum in developing an efficient long-term conservation policy their presence may mitigate deforestation effects for many species; this also applies in tropical regions (Turner and Corlett 1996), where fragmentation is more deleterious than in temperate and boreal forests owing to the greater intolerance of species to habitat change.

Several types of disturbance occur in fragments of tropical forests. Some effects are common to other forests, for example the human disturbance of harvesting, or the diversion of watercourses for transporting timber. Other effects are particularly evident in these fragments, for example the edge effect, the change of microclimate, making it drier and hotter at the border. The loss of keystone species like large carnivores causes an increase in some species of small carnivores, which in turn creates pressure on prey species. Few rainforest species are tolerant of open spaces, and fragmentation increases the number of alien species that compete with the indigenous species.

Despite these negative effects fragments are better than nothing (Turner and Corlett 1996). The presence of tropical fragments in an agricultural landscape greatly enhances biological diversity (Nepstad *et al.* 1996). A realistic policy in rural Amazonia should ensure forest remnants: this is a very important step (Schelhas and Greenberg 1996). To increase this capacity it is important to ensure greater connectedness between patches. Fragments in many cases are enough to preserve invertebrate fauna and to ensure the spreading of plants.

We can expect a species decline over time in small fragments, but these fragments could provide opportunities to preserve species from definitive extinction in the future. The presence of fragments could recreate in the future, by coalescence, a new tropical forest cover, a process which is difficult when starting from bare soil.

Cardoso da Silva *et al.* (1996) suggest mitigation action during tropical forest clearing by reducing the size of clearcut pastures in order to increase the colonization capacity of forest trees after 6–8 years of abandonment. They observe that birds, especially frugivorous birds, crossing the pastures can spread tree seeds, thereby favouring forest recovery. *Ad hoc* legislation on the size and shape of deforested strips could strongly mitigate the effect of forest fragmentation (Fig. 7.3).

Most actual fragmentation in the tropics is driven by short-term economic constraints and solutions to reverse this trend are not on the agenda.

Table 7.2 lists herd sizes in the northern region of Brazil. The huge increase in deforested areas is evident.

Pastures in tropical lands are not self-maintained as they are in temperate biomes. In the tropics most nutrient cycles are maintained in the vegetation biomass. The soil has little capacity to retain nutrients, so after a few years the pastures are impoverished and the grasses substituted by unpalatable shrubs. Removing the shrubs and adding fertilizer is possible but inconvenient, and new pastures are created by deforestation, in a type of shifting ranching (Hecht 1993). Most pasture land older than 10 years is abandoned and transformed into degraded land.

In Australia the maintenance of non-logged creeks and gullies seems a good strategy to conserve nocturnal birds and mammals (Kavanagh and Bamkin 1995).

Silviculture in the eucalyptus forests of Tasmania has been discussed in terms of bird species conservation. Logging retention seems a promising technique to reduce the impact of logging on animal communities. Tree retention produces a lower density of species but maintains high diversity compared to unlogged stands in bird assemblages (Taylor and Haseler 1995) (Fig. 7.4).

Temperate regions support fragmentation better than tropical ones, having species more adapted to fragmentation constraints. Villard *et al.* (1995), in a fragmented rural area of Ontario, found that the population turnover of four target species of neotropical migratory birds had was due to a combination of many factors, and that site fidelity was

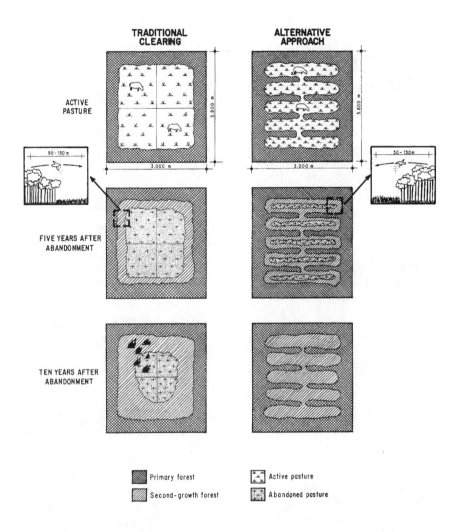

Figure 7.3 Schematic representation of traditional clearcutting in Amazonia compared with the recommended method of opening small strips of forest connected with each other, large enough to assure grazing efficiency but early overflown by seed-disseminating birds (from Cardoso da Silva *et al.* 1996, with permission).

Table 7.2 Herd size in north Brazil in the period 1970–85. It is possible to verify that cows have a dramatic increase in most of the regions. These values may be used to evaluate the level of fragmentation of tropical forest in this country (from Hecht 1993, with permission).

	1970	1975	1980	1985	% increase 1970–1985
Acre	72,166	120,143	292,191	333,457	362
Amazonas	283,362	415,457	455,584	420,940	48
Para	594,313	777,660	2,729,796	3,485,368	486
Rondonia	23,126	55,392	248,558	768,411	3227
Roraima	238,761	246,126	313,069	303,501	27
Amapa	64,990	62,660	46,069	46,901	−38

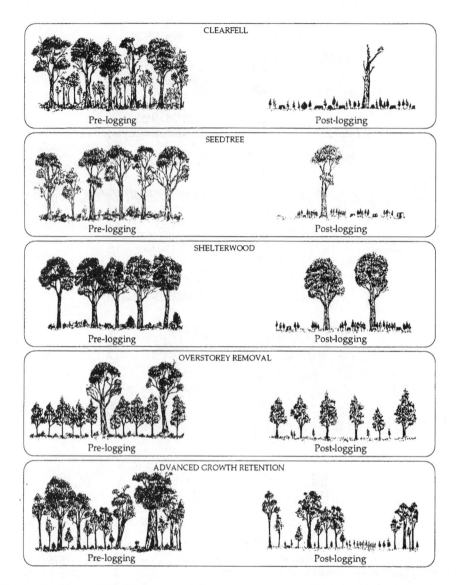

Figure 7.4 Silvicultural treatments. Clearfell: removal of all trees; seedtree retention: 7–15 well-spaced trees per hectare; shelterwood: logging is reduced to ensure a good soil cover; overstorey removal: logging of old trees in a two-aged stand, reducing competition between two strata; advanced growth retention: in a multiaged structured forest. All trees with good growth potential are retained (from Taylor and Haseler 1995, with permission).

an important element. This research contributes to a long debate on the role of island size and isolation. The behaviour of species can change most of our actual knowledge.

Isolated woodlots are important when animals such as migratory birds are moving at landscape scale. In east-central Illinois Blacke (1986) found that small woodlots are important for migratory birds, especially if these elements are surrounded by large forested areas. The presence of many species during migration confirms the ability of birds to colonize fragmented habitats, but probably other negative factors, such as high predatory pressure, no tolerance of edges, a more unpredictable microclimate and higher intra- and interspecific competition contribute to local rarefaction.

7.6.3 Conserving large carnivores

The conservation of large carnivores poses more problems in any conservation plan because the minimum area required by these mammals generally surpasses the size of the reserve (Noss *et al.* 1996). The traditional method of selecting discrete nature reserves generally is not enough to protect the carnivore populations efficiently, and these authors

suggest a zoning approach in which core reserves are surrounded by a buffer and connected by regional corridors to other reserves (Fig. 7.5).

This model has three main elements: a core area, which is a large, wild, low road-density area that should play a fundamental source role; the second element is a buffer zone or zone of transition, which can be extremely important for large home-range species. In this area human activity is reduced and a political and social compromise ensures only modest pressure on large carnivores. Thirdly are the corridors that should assure the movement of animals

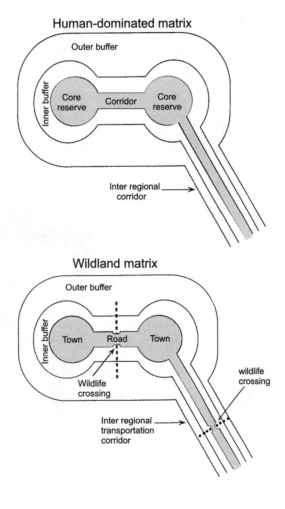

Figure 7.5 Schematic example of a human-dominated region (top) consisting of core reserves connected by corridors and surrounded by a buffer zone that mitigates the external, negative influence of highly populated areas. In the bottom case a wildland matrix incorporates a town and road system; the connectivity of the wild area is assured by wildlife facilities such as tunnels and overpasses (from Noss *et al.* 1996, with permission).

to guarantee genetic exchange between the metapopulations living in surrounding reserves. This model has general potential and can be used in all human-dominated landscapes at different spatial scales.

It appears important to take into account also the scale at which we consider connectivity. In fact conservation planners should consider the movement of individuals within home ranges, and dispersal between home ranges or between populations. The source–sink model, in terms of the quality of habitat considered, is extremely important for planning efficient reserves.

Corridors again have not been considered fixed structures of a landscape but a functioning part of the landscape that change according to season, species and landscape disturbance. In some cases the corridors preferred by large carnivores are the same parts of the landscape preferred by humans.

Natural reserves considered as discrete and isolated entities in a human-dominated landscape cannot be a successful model to protect animals such as large carnivores. The need to manage the regional landscape seems a more promising approach in human-dominated systems.

In Italy the recent spread of the wolf (*Canis lupus*) is a clear example that if connectivity is assured (in this case the land abandonment along the Apennines) a huge system of secondary succession core areas is created that assures refuges and movement to this species. At the same time, because of a decrease in livestock (the main food resource for wolves in the past) owing to land abandonment, deer and wild boar have recolonized most of the mountain ranges from the Alps to the southern Apennines, Sardinia and Sicily, producing new food for the wolf (Fig. 7.6).

7.6.4 Toward the conservation of processes

Conservation plans should take into account the maintenance of ecological fluxes rather than focusing on the conservation of species. Often the presence of a species is ephemeral, linked to a particular stage of ecological succession.

In Europe an important goal should be the conservation of bird migratory processes. At each season millions of birds move across Europe to the Mediterranean and Africa using different movement strategies, from no-stop flight to stopover flights

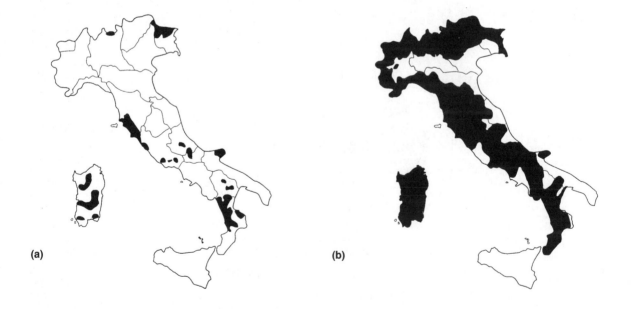

(a)

(b)

Figure 7.6 Distribution of ungulates (*Cervus elaphus, Dama dama, Capreolus capreolus, Ovis musimon, Sus scrofa*) at the beginning of this century (**a**) and the present distribution (**b**) (**a** from Ghigi (1911) and **b** from Pavan (1983), quoted by Apollonio 1996, with permission).

(Fig. 7.7). In particular, birds belonging to the second group require specific habitats in which to roost, to refresh and to forage in order to replace the energy lost in migration. Generally for such birds open, park-like areas in which short grass patches are blended with shrubland and woodland, are suitable temporary habitats.

Many species are habitat specialists during the breeding season, but during migration they behave as habitat opportunists. Large-scale migratory refuges are necessary to maintain these fluxes and this strategy requires strong political consensus among different countries. In this case farmland and pasture land should be considered with priority as valuable landscapes to preserve the migratory birds of west-palearctic regions.

At the other end of the spectrum the conservation of less vagile species requires strategies that take into account the rehabilitation and the creation of new source habitat patches for target species. These strategies are not in conflict but are concurrent.

7.6.5 Landscape patterns and conservation

Landscape patterns are represented by the distribution in space and time of habitat patches and resources. Fahrig and Merriam (1994) have divided landscape patterns into two broad categories, spatial and spatiotemporal. In the first case time is not considered important, at least at the scale of the species investigated. In the second case the landscape pattern changes in time and space, for example a habitat, can be ephemeral.

Harrison and Fahrig (1995) consider six main components of the landscape spatial pattern:

- Amount of habitat in the landscape;
- Mean size of habitat patches;
- Mean interpatch distance;
- Variations in patch size;
- Variations in interpatch distance;
- Landscape connectivity.

Figure 7.8 illustrates the increasing probability of population survival at a regional scale.

Figure 7.7 Abandoned pastures in southern Tuscany landscape, preferred habitats for stopover migrant birds that find available resources to replenish energy to complete the migratory trip from Africa to central-north Europe.

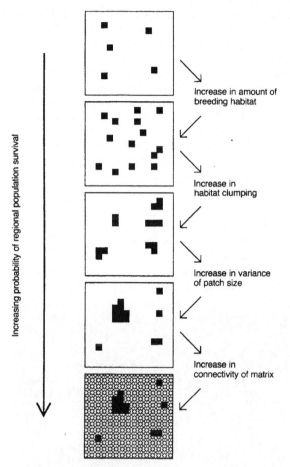

Figure 7.8 Predictive modelling of the effects of the landscape spatial pattern on regional population survival (from Harrison and Fahrig 1995, with permission).

If the temporal dimension is introduced into a landscape spatial pattern we can expect changes in that pattern over time, and in the end the survival of a species is affected. Disturbance and permanence of a temporary or ephemeral patch are important factors. Disturbance can alter survival according to rate, size and temporal correlation with disturbance. The rate of patch formation and patch lifespan are important aspects of ephemeral patches.

There is a general consensus that habitat loss (e.g. due to fragmentation) reduces the probability of survival of a regional population, but if patch size increases and the interpatch variance in patch size also increases, this has a beneficial effect on population survival.

In an ephemeral habitat the survival probability increases if the patch lifespan increases, and in

such types of landscape the spatial arrangement of the patches is less important, owing to the high dynamics.

7.7 LANDSCAPE DESIGN (CREATION) AND RESTORATION

A main goal of restoration in landscape ecology is to identify key habitats and species, to assess distributional gaps and to process the best strategies for plants and wildlife. At the same time recreational benefits to society must be guaranteed (Anderson 1995). Unfortunately, restoration projects for large areas are rare, and this is an intrinsic limitation in the application of landscape ecology principles. But in some cases, as with the restoration and design of ponds, their presence in clusters could provide more opportunities for metapopulations than a few large isolated areas.

Landscape design is an important component of practical landscape ecology. Generally landscape design is considered an activity to rehabilitate degraded landscapes, or modify landscapes after a change in land use, as with fields in urban fringes. This important branch of landscape ecology is too far reaching to include in this book, but some guidelines will be given. The principle that guides landscape design is, in summary, the synthesis between human perception and ecological processes. Modern landscape design sees visual patterns as the prime criterion, creating a harmonious view of neighbouring areas. This depends largely on the countryside-oriented sensitivity that creates recognizable British, central European or Mediterranean landscapes.

Landscape design in most cases consists in planting trees and shrubs in appropriate quantities, shapes and diversity, mimicking natural patterns and increasing visual and structural complexity. These actions in general are favourable to many species of plants and animals, and produce new ecological processes.

Designing a new landscape is not a common activity but it is very popular to restore small areas used in the past as gravel pits or for industrial waste disposal.

The variety of local conditions prevents a detailed discussion, but some general guidelines can be listed:

- Select indigenous species of trees and shrubs;
- Plant according to natural shape and not in linear rows, and never perpendicular to the contour;
- Increase the edge shape complexity, modulated on minor soil undulations;

- Retain trees wherever possible. Scattered trees are important for wild fauna;
- Planting design should pay attention to the different habitat needs of target species in terms of soil, sunlight, sheltered or exposed sites.

Landscape creation is a relatively recent practice, although the cultural landscapes are beautiful examples from the past. However, a fundamental difference appears when we compare the landscapes of the past and the present. In the past the landscape reaction was dominated by a balance between human input and land productivity by a continuous stewardship. Today most landscape management activity focuses on wildlife conservation and scenic improvement.

Many tools are available in ecological restoration, and it is common to reinstate traditional management where possible, but in many cases this is too expensive for managing entire landscapes. Grazing, mowing and fire are common practices at local as well as at landscape scale. Grazing is particularly efficient and a relatively cheap method that can be used from lowlands to uplands. Fire is very efficient but not usable in every condition and season, and is more difficult to control when dry biomass has accumulated.

The deciduous forests of the eastern United States have been reduced in a few generations to a very low level. In order to guarantee the survival of these stands for future generations, Keddy and Drummond (1996) proposed a four-step approach:

1. Manage remaining forest areas sustainably.
2. Restore altered forests to their original composition.
3. Replant deforested areas.
4. Protect remnant primeval areas as an ecological model for research and comparison.

To achieve these goals some descriptors were presented at stand scale and at landscape scale (Table 7.3).

7.8 HIERARCHICAL STRUCTURE OF THE SYSTEM AND CONSERVATION OF BIODIVERSITY

Conservation strategies are moving from a static species- or habitat-oriented policy to a more dynamic and realistic approach. In this direction a fundamental contribution is made by hierarchy theory. The hierarchical approach to understanding natural systems, as described in Chapter 2, allows us to take into consideration the different component subsystems, assuming a different dynamic moving from the smallest (faster dynamic) to the largest (low dynamic) (Lewis *et al.* 1996). The aim of this approach is to protect the total diversity at the landscape level of ecological organization (Norton and Ulanowicz 1992) and to distinguish different policies according to the hierarchical scale selected.

Before explaining these concepts it is important to focus on the autopoietic capacity of the natural system to be creative. At every level, systems have the capacity to self-sustain through homoeostatic and homeorhetic responses to changing conditions. This ensures the system's ability to adapt to new conditions, and may be considered a sign of a healthy ecological system.

Landscape scale seems essential in any biodiversity protection plan, overturning the myth that species conservation must be a good start at the small and medium scale. Biological and ecological (*sensu* Naveh 1994) diversity should be considered 'dynamically in terms of healthy processes, rather than merely as maintenance of current elements of the system' (Norton and Ulanowicz 1992).

In this perspective economy and diversity should find a meeting point. In fact, in a world dominated by humans we need to discover mechanisms that enhance diversity, encouraging policies that in the final analysis mimic the natural disturbance regime (Fig. 7.9). We can imagine the socioeconomic evolution moving in parallel with the ecological one and presenting a trajectory shaped by reciprocal feedback.

The socioeconomic crisis of humanity in different times and in different parts of the world has always been followed by a divergence from the processes of ecological systems. We can learn an important lesson from the Everglades. This system has been managed in different ways according to the different crises that have created the conditions for new management (Gunderson *et al.* 1995) (Fig. 7.10).

Including the spatial dimension in ecological conservation means taking into account variables often neglected, and spatial effects that can have consequences over time. This is the case with the hypothesis presented by Tilman *et al.* (1994) on the behaviour of species extinction moving with habitat destruction. The more a habitat is fragmented the higher the number of extinctions caused by extra destruction. This effect, occurring generations after fragmentation, represents a future cost of current habitat destruction (Fig. 7.11).

Table 7.3   Main ecological descriptors used in deciduous forest in eastern United States by Keddy and Drummond (1996) at stand (local) and landscape scale (from Keddy and Drummond 1996, with permission)

Stand scale	1. Tree size defined in terms of basal area (mean diameter at breast height (dbh))	Three categories : normal ($>$29), intermediate (20–29) and low ($<$20 m²/ha)
	2. Canopy composition	In a mature deciduous forest, the canopy is dominated by a few species, while young forests have a higher tree diversity. In a mature undisturbed forest, shade-tolerant species are dominant and intolerant occupy 2–10% of total forest composition. On increasing the disturbance, shade-tolerant species increase. Hence the percentage of tolerant species can represent a property. They distinguish three categories: control/normal ($>$70% of tolerant species), intermediate (30–70%) and heavily altered ($<$30%)
	3. Coarse woody debris	This material, which includes fallen logs, snags and large branches, is important for many species of organisms. The coarse woody debris in a forest depends on the history of the forest and generally increases with age. Three categories are proposed: control/normal (both firm and crumbling large log present), intermediate (either firm or crumbling large logs present) and low (large logs absent)
	4. Herbaceous layer	Herbs are present in mature forest in gaps and clearings. This layer is particularly sensitive to grazing. Three categories are suggested: control/normal ($\geq$6 species), intermediate (2–5 species) and low ($<$2 species)
	5. Curticulous bryophytes (mosses. liverworts and lichens)	Their presence in many cases enriches nitrogen in the soil. Three categories are proposed: control/normal ($\geq$7 species), intermediate (2–6 species) and low ($<$2 species)
	6. Wildlife trees	The presence of large logs and trees with cavities enhances the suitability of the forest for animals (especially mammals and birds). Three categories are proposed: control/normal ($\geq$3 wildlife trees per 10 ha), intermediate (1–3 wildlife trees per 10 ha), low ($<$1 wildlife tree per 10 ha)
	7. Fungi: macrofungi	No quantitative data available. Probably a good indicator to develop in the future
Landscape scale	8. Avian community	Many birds are not sensitive to forest age, and often there are more species in intermediate-aged stands than in old-growth forest, but some species require a large forested area. These area-sensitive species may be good indicators of stand size. Three categories are proposed: control/normal ($\geq$5 area sensitive species), intermediate (2–4 species), and low ($<$2 species)
	9. Large carnivores	Generally undisturbed forest have a high number of carnivores such as eastern cougar, black bear, fisher, bobcat, wolf and fox. Three categories are suggested: control/normal ($\geq$6 species), intermediate (3–5 species), low ($<$3 species)
	10. Forest area	This descriptor is very important. Disturbed forests are generally fragmented into woodlots. In order to maintain an avian community typical of undisturbed stands 75 ha of continuous forest are necessary and 100 000 ha for a complete assemblage of mammals. Three categories are proposed: control/normal ($>10^6$ ha), intermediate (10^2–10^6 ha), low ($<10^2$ ha)

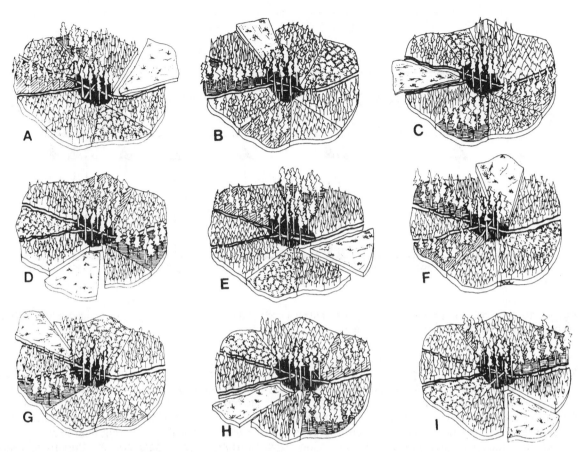

Figure 7.9 Long-term logging rotation to mitigate fragmentation in old-growth forests. This system should ensure that almost 66% of the surrounding forested buffer is more than 100 years old. The 33% regeneration area will provide foraging and habitat for successional species (from Harris 1984, with permission).

7.9 SUMMARY

- Landscape ecology represents a scientific basis for studying, planning and managing seminatural, rural and agricultural landscapes.
- Landscape evaluation is an important step in preparing significant actions toward the management of large areas, taking into consideration natural and human-induced processes.
- In a world largely dominated by humans the concept of a cultural landscape for many ecological systems appears extremely useful in understanding the complexity of such systems.
- The cultural landscapes are the product of long-term interactions (feedback) between natural and human processes.

- Recently Unesco and IUCN have recognized the importance of preserving valuable cultural landscapes around the world.
- Most of the cultural landscapes have high biological and ecological diversity but require constant human stewardship for their maintenance.
- The unequal distribution of energy, resources and organisms requires *ad hoc* tools for nondestructive management.
- The application of the principles of landscape ecology to management is full of promise. In fact, the landscape scale is comprehensive of socioeconomic and natural processes.
- Often the administrative boundaries of land division overlap ecological entities as recognized by the landscape classification.

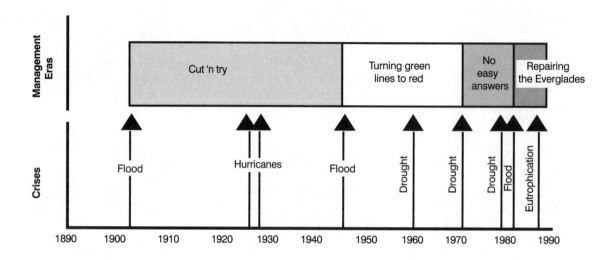

Figure 7.10 Historical process shared between environmental crisis and management policy in the Everglades from 1980 to the present (from Gunderson *et al.* 1995, with permission).

• Moving from theory into practice allows us to apply the principle of landscape ecology to design and management of nature reserves.

• During recent decades nature conservation and landscape ecology have found common points, and the strategy to give priority to conserving processes rather than ephemeral patterns seems very appropriate.

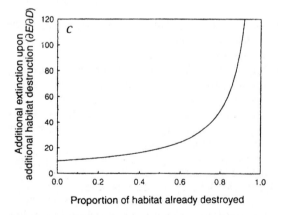

Figure 7.11 Representation of the pattern of additional extinction in habitat already destroyed (from Tilman *et al.* 1994, with permission).

• The maintenance of the natural disturbance regime seems a promising approach to perpetuate biological and ecological diversity. However, this approach is often in conflict with the size of reserves and with conflicting land uses.

• Corridors and hedgerows are important structures to be conserved in the landscape, ensuring movement and dispersal to many organisms. Their maintenance is often in conflict with socioeconomic development policies.

• Owing to worldwide human intrusion in natural systems habitat fragmentation seems inevitable. In modified landscapes such as intensive-rural landscapes a small fragment of natural forest can conserve a valuable biodiversity.

• Landscape ecology resurrects the value and the importance of such fragments, producing guidelines for profitable management.

SUGGESTED READING

Harris, L.D. *The fragmented forest. Island biogeography theory and the preservation of biotic diversity.* University of Chicago Press, Chicago, 1984.

Westman, W.E. *Ecology, impact assessment and environmental planning.* John Wiley and Sons, New York, 1985.

Berger, J.J. (ed.) *Environmental restoration. Science and strategies for restoring the earth.* Island Press, Washington, DC, 1990.

Community woodland design. Guidelines. HMSO, London, 1991.

Lowland landscape design. Guidelines. HMSO, London, 1992.

Austad, I., Hauge, L., Helle, T. *Maintenance and conservation of the cultural landscape in Sogn og Fjordane, Norway.* Department of Landscape Ecology, Sogn og Fjordane College, Norway, 1993.

von Droste, B., Plachter, H., Rossler, M. (eds.) *Cultural landscapes of universal value.* Gustav Fischer, Jena, 1995.

Halladay, D. and Gilmour, D.A. (eds.) *Conservation biodiversity outside protected areas. The role of traditional agro-ecosystems.* IUCN (Gland), Switzerland, and Cambridge, 1995.

REFERENCES

Anderson, J.E. (1991) A conceptual framework for evaluating and quantifying naturalness. *Conservation Biology* **5**: 347–352.

Anderson, P. (1995) Ecological restoration and creation: a review. *Biological Journal of the Linnaean Society* **56** (suppl.): 187–211.

Apollonio, M. (1996) Evoluzione dell'ambiente e delle attività antropiche nelle aree appenniniche in relazione alla presenza del lupo (*Canis lupus*). In: Cecere, F. (ed.), *Atti del Convegno Nazionale Dalla parte del lupo.* Atti e Studi del WWF Italia, no. 10, pp. 54–63.

Baker, W.L. (1992) The landscape ecology of large disturbances in the design and management of nature reserves. *Landscape Ecology* **7**: 181–194.

Blake, J.G. (1986) Species–area relationship of migrants in isolated woodlots in east-central Illinois. *Wilson Bulletin* **98**: 291–296.

Burel, F. (1996) Hedgerows and their role in agricultural landscapes. *Critical Reviews in Plant Sciences* **15**: 169–190.

Cardoso da Silva, J.M., Uhl, C., Murray, G. (1996) Plant succession, landscape management, and the ecology of frugivorous birds in abandoned Amazonian pastures. *Conservation Biology* **10**: 491–503.

Fahrig, L. and Merriam, G. (1994) Conservation of fragmented populations. *Conservation Biology* **8**: 50–59.

Farina, A. (1995) Cultural landscapes and fauna. In: von Droste, B., Plachter, H., Rossler, M. (eds.) *Cultural landscapes of universal value.* Gustav Fischer, Jena, pp. 60–77.

Farina, A. (1996) The cultural landscape of Lunigiana. *Memorie Accademia Lunigianese Scienze Giovanni Capellini* **LXVI**: 83–90.

Gotzmark, F. (1992) Naturalness as an evaluation criterion in nature conservation: a response to Anderson. *Conservation Biology* **6**: 455–460

Gunderson, L.H., Light, S.S., Holling, C.S. (1995) Lessons from the Everglades. *Bioscience* Supplement: S-66–73.

Haber, W. (1990) Using landscape ecology in planning and management. In: Zonneveld, I.S. and Forman, R.T.T. (eds.), *Changing landscapes: an ecological perspective.* Springer-Verlag, New York, pp. 217–232.

Halladay, D. and Gilmour, D.A. (eds.) (1995) *Conservation biodiversity outside protected areas. The role of traditional agro-ecosystems.* IUCN (Gland), Switzerland, and Cambridge, UK.

Hansson, L. and Angelstam, P. (1991) Landscape ecology as a theoretical basis for nature conservation. *Landscape Ecology* **5**: 191–201.

Harris, L.D. (1984) *The fragmented forest. Island biogeography theory and the preservation of biotic diversity.* University of Chicago Press, Chicago.

Harrison, S. and Fahrig, L. (1995) Landscape pattern and population conservation. In: Hansson, L., Fahrig, L., Merriam, G. (eds.), *Mosaic landscapes and ecological processes*, Chapman & Hall, London, pp.294–308.

Harrison, R.L. (1992) Toward a theory of inter-refuge corridor design. *Conservation Biology* **6**: 293–295.

Hecht, S.B. (1993) The logic of livestock and deforestation in Amazonia. *BioScience* **43**: 687–695.

Jorg, E. (1994) *Field margin-strip programmes.* Proceedings of a technical seminar organized by the Landesanstalt für Pflanzenbau und Pflanzenschutz, Mainz.

Kavanagh, R.P. and Bamkin, K.L. (1995) Distribution of nocturnal forest birds and mammals in relation to the logging mosaic in southeastern New South Wales, Australia. *Biological Conservation* **71**: 41–53.

Keddy, P.A. and Drummond, C.G. (1996) Ecological properties for the evaluation, management, and restoration of temperate deciduous forest ecosystems. *Ecological Applications* **6**: 748–762.

Lewis, C.A., Lester, N.P., Bradshaw, A.D. *et al.* (1996) Consideration of scale in habitat conservation and restoration. *Canadian Journal of Fisheries and Aquatic Science* **53**: 440–445.

Lindemayer, D.B. and Nix, H.A. (1993) Ecological principles for the design of wildlife corridors. *Conservation Biology* **7**: 627–630.

Mann, C.C. and Plummer, M.L. (1993) The high cost of biodiversity. *Science* **260**: 1868–1871.

Mills, L.S., Soul,, M.E., Doak, D.F. (1993) The keystone-species concept in ecology and conservation. *Bioscience* **43**: 219–224.

Naveh, Z. (1994) From biodiversity to ecodiversity: a landscape-ecology approach to conservation and restoration. *Restoration Ecology* **2**: 180–189.

Nepstad, D.C., Moutinho, P.R., Uhl, C., Vieira, I.C., da Silva,C. J.M. (1996) The ecological importance of forest remnants in an eastern Amazonian frontier landscape. In: Schelhas, J. and Greenberg, R. (eds.) *Forest patches in tropical landscapes.* Island Press, Washington DC, pp. 133–149.

Norton, B.G. and Ulanowicz, R.E. (1992) Scale and biological policy: a hierarchical approach. *Ambio* **21**: 244–249.

Noss, R.F. (1995) Foreword. In: Hansson, L., Fahrig, I., Merriam,G. (eds.), *Mosaic landscapes and ecological processes.* Chapman & Hall, London.

Noss, R.F., Quigley, H.B., Hornocker, M.G., Merrill, T., Paquet, P.C. (1996) Conservation biology and carnivores conservation in the Rocky mountains. *Conservation Biology* **10**: 949–963.

Paine, R.T. (1966) Food web complexity and species diversity. *American Naturalist* **100**: 65–75.

Paine, R.T. (1969) A note on trophic complexity and community stability. *American Naturalist* **103**: 91–93.

Plachter, H. and Rossler, M. (1995) Cultural landscapes: reconnecting culture and nature. In: van Droste, B., Plachter, H., Rossler, M. (eds.) *Cultural landscapes of universal value.* Gustav Fischer, Jena.

Schelhas, J. and Greenberg, R. (eds.) (1996) *Forest patches in tropical landscapes.* Island Press, Washington.

Shafer, C.L. (1995) Values and shortcomings of small reserves. *Bioscience* **45**: 80–88.

Taylor, R.J. and Haseler, M.E. (1995) Effects of partial logging system on bird assemblages in Tasmania. *Forest Ecology and Management* **72**: 131–149.

Tilman, D., May, R.M., Lehman, C.L., Nowak, M.A. (1994) Habitat destruction and the extinction debt. *Nature* **371**: 65–66.

Turner, I.M. and Corlett, R.T. (1996) The conservation value of small, isolated fragments of lowland tropical rain forest. *TREE* **11**:330–333.

Turner, M.G., Arthaud, G.J., Engstrom, R.T. *et al.* (1995) Usefulness of spatially explicit population models in land management. *Ecological Applications* **5**: 12–16.

von Droste, B., Plachter, H., Rossler, M. (eds.) (1995) *Cultural landscapes of universal value.* Gustav Fischer, Jena.

van Wilgen, B.W., Cowlig, R.M., Burgers, C.J. (1996) Valuation of ecosystem services. A case study from South African fymbos ecosystems. *Bioscience* **46**: 184–189.

Villard, M., Merriam, G., Maurer, B.A. (1995) Dynamics in subdivided populations of neotropical migratory birds in a fragmented temperate forest. *Ecology* **76**: 27–40.

Wilcove, D.S. (1989) Protecting biodiversity in multiple-use lands: lessons from US Forest Service. *TREE* **4**: 385–388.

Methods in landscape ecology 8

8.1 INTRODUCTION

This chapter is particularly important because it is an attempt to quantify, using different approaches, the main attributes of a landscape. Routines and practical examples are provided to guide the reader through a number of ways to quantify many but not all the attributes of landscapes.

Landscape approaches are so diverse that it is not possible to review them all comprehensively and to indicate standard methodologies. Many come from geostatistics, geobotanics, animal population analysis, behavioural ecology etc. Image processing, geographic information systems, spatial statistics and fractal geometry represent the most common ways to explain landscape complexity, but there are many others.

Generally maps, aerial photographs and satellite images are commonly used before and after any field inventory. This material suffers from many biases due to time, resolution and quality, and often it is difficult to normalize the information that can be drawn.

Depending on the scale of resolution we need, we move from a coarse-grained resolution of 30 m or more provided by Landsat images, to the 10 m resolution of Spot images of satellites. Using aerial photographs the resolution may range from 2–3 m to a few centimetres, depending on the altitude at which the images were obtained.

Recently the use of video cameras suspended from a low-altitude balloon has opened new perspectives in remote sensing techniques. For instance, a video camera has been utilized to study the evolution of vegetation on mountain prairies. Some examples are presented in this chapter but most of the data are still under study.

Collecting field data requires us to know exactly where we are, and so the use of detailed maps (1:5000 to 1:25 000) is required. Recently we have been able to localize features of interest using the Global Positioning System (GPS). This system, developed for missile and aeroplane automatic navigation systems, is based on radio information on the field position calculated by a cluster of satellites. A practical use of GPS has been in the study of bird communities (Farina 1997). The birds' positions were entered by an operator into a data logger and transferred to a computer for processing. This information was then inserted into a GIS for mapping and spatiostatistical elaboration.

For the most popular measures we have prepared simple and unsophisticated programs in Basic language, reported in the Appendix.

8.2 NUMERICAL AND SPATIAL DATA PROCESSING

The spatial elaboration of data is central to landscape ecology, and for this reason a large amount of space is dedicated to this argument. Many techniques have been borrowed, mainly from spatial statistics or geostatistics, image analysis and fractal geometry. Euclidean and non-Euclidean geometry are often combined to analyse the complexity of spatial processes and patterns across temporal and spatial scales.

Two types of information are processed in this analysis: the patch attributes and the landscape attributes. In the first case the analysis focuses on every patch in terms of size, shape and spatial arrangement. In the second case more complicated analyses are necessary to explore the complexity of the land mosaic. For the second approach in particular, a great variability can be expected on changing the scale of resolution.

8.2.1 Measures of patch characteristics (see Basic routine in Box 8.1)

Patch size S is the measure of the area of each patch composing a mosaic (Fig. 8.1).

Patch Perimeter L is the measure of the perimeter of each patch composing a mosaic.

Patch shape Many indices exist to measure patch shape, especially in a geographical context. We have selected some used in direct landscape analysis. These indices must be adopted and used with caution because often their precise ecological meaning is not so easily found. The approach to the study of patch shape is important for the consequences that patch regularity/irregularity has on organisms. We assume a circle to be a regular patch; the more irregular a patch the more edges and less

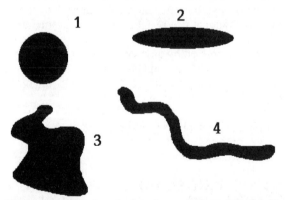

Figure 8.1 Examples of patch shape which can be found in a landscape: from regular close to a circle (no. 1) of anthropogenic origin, to a strip belt (no. 4) more related to natural events such as stream corridors. See Table 8.1 for geometric attributes.

interior area are available. An irregular patch probably has more heterogeneous processes than a regular one. Habitat suitability, predation risk and microclimatic stress are some of the direct consequences of an irregular patch. This, of course, is important for some species but not for all. Six indices for calculating patch shape are described:

1. **Perimeter–area ratio** (*L/S*)
   The perimeter of each patch is divided by its area: *L/S* where *L* = perimeter and *S* = area. This index varies according to the size of the patch even when the shape is constant. See Buechner (1989) for an application to the field study of mammal dispersal.

2. **Corrected perimeter–area (CPA)**
   This index is corrected for solving the size problems of index 1 and varies between 0.0, a perfect circle, and infinity for an infinitely long and narrow shape:
   $CPA = (0.282 \times L)/\sqrt{S}$

3. **Related circumscribing circle (RCC)**
   This index compares the patch size with the size of a circle that can circumscribe the patch:
   $RCC = 2*(area/\pi)^{1/2}/longest\ axis$
   This index varies between 0.0 and 1.0 as the shape of the patch approaches a circle.

4. **S1** (Hulshoff 1995)
   $S1 = 1/Ni*\Sigma(Li/Si)$ where *Ni* is the number of patches of category *i* in a map, *Li* is the perimeter and *Si* the area of each patch in category *i*. A high value of this index indicates the presence of many patches with small interiors.

5. **S2** (Hulshoff 1995) is the measure of isodiametric attributes of patches:
   $S2 = 1/Ni*\Sigma(Li/4\sqrt{Si})$ where *Ni* is the number of patches of category *i*, *Li* is the perimeter and *Si* is the size of each patch in the category. The further *S2* is far from 1, the more the patches deviate from an isodiametric shape.

6. **Fractal dimension D** (see more on fractals in Section 8.3). The complexity of path shape can be measured by regressing the log of patch perimeter *L* with the log of patch size *S*:
   $D = 2s$
   where *s* is the slope of the regression.

8.2.2 Measure of the spatial arrangement of patches

Richness is the number of different patch attributes that are present in the study area.

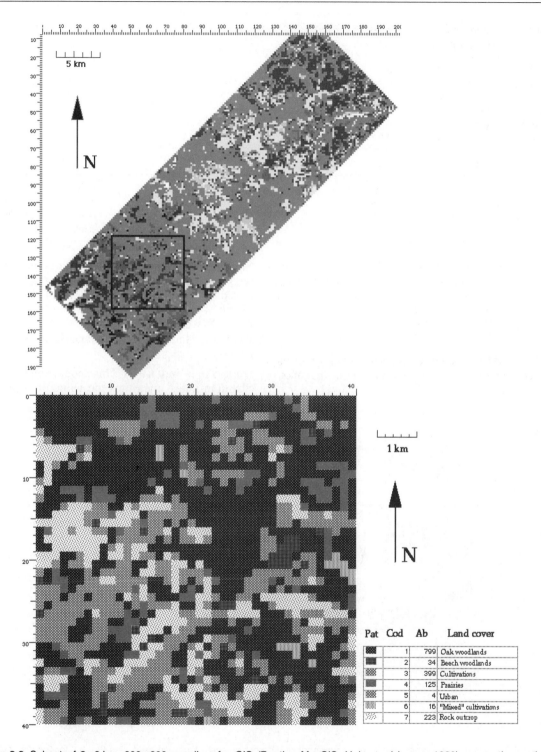

Pat	Cod	Ab	Land cover
	1	799	Oak woodlands
	2	34	Beech woodlands
	3	399	Cultivations
	4	125	Prairies
	5	4	Urban
	6	16	"Mixed" cultivations
	7	223	Rock outcrop

Figure 8.2 Subset of 8×8 km, 200×200 m cells, of a GIS (Routine MacGIS; Hulse and Larsen 1989) across the northern Apennines (from Nardelli 1995, with permission). The mosaic is composed of seven main land cover types. Pat = graphic pattern; Cod = land cover code; AB = number of cells.

Table 8.1 Some measures of patches as in Fig. 8.1.
#patch = patch identifier;
S = area (patch size, in pixels);
L = perimeter (patch edge, in pixels);
L/S = perimeter/area ratio;
Corr (corrected perimeter–area) = 0.282*L/√S;
RCC (related circumscribing circle) = 2*(S/π)^½/longest axis;
Long axis (longest axis).
Data were processed using the routine in Basic of Box 8.1.

#patch	S	L	L/S	CPA	RCC	Long axis
1	4838	259.52	0.054	1.052	0.994	79.010
2	4235	356.27	0.084	1.544	0.451	162.720
3	11055	521.24	0.047	1.398	0.867	136.900
4	5639	662.32	0.117	2.487	0.448	189.090

Number of patches is the number of patches present in a mosaic. This index can be cumulative for all the mosaic or for each land cover or vegetation type, and is particularly sensitive to rare land cover (Figs 8.2, 8.3 and Table 8.1).

Relative abundance is a measure of the proportion of each land cover or vegetation type in the study area.

Shannon diversity (*H'*) This index combines richness and evenness. The variety and relative abundance of land cover can be estimated using the Shannon index:

$$H' = -\Sigma Pi \ln Pi$$

where *Pi* is the relative importance of the land cover of type *i*.

Dominance index (*D*) This index, related to the Shannon index, measures the value of dominance

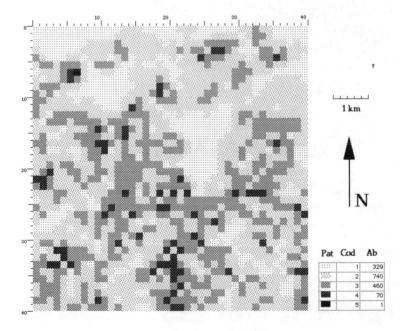

Figure 8.3 Distribution of edges (ecotones) as number of different adjacent land use cells. See Fig. 8.2 for the display of the land cover. Edges have been calculated using the routine scan-diversity from MacGIS (Hulse and Larsen 1989). Pat = graphic pattern; Cod = edge code (1 = no edge, 2 = two land covers, 3 = three land covers, 4 = four land covers, 5 = five land covers round each cell at a distance of one cell); Ab = number of cells for each category, see Cod.

of one land cover over the others (O'Neill *et al.* 1988):

$$D = \ln n - H'$$

where $H' = -\Sigma Pi \ln Pi$, where Pi is the proportion of the grid cells on the landscape for the land use i selected; n is the number of land use categories. D is close to 0 when the land cover types present an equiabundance and is close to 1 when most of the land cover belongs to one cover type. In our example $D = 0.464$.

Evenness (E) This index responds to the number of patch types and their proportion in the landscape:

$$E = -100 \ln(\Sigma Pi^2)/\ln(n)$$

where n is the total number of patch types and Pi is the probability that a pixel belongs to type i. See Romme (1982) for an example of application.

Relative patchiness (RPI)

$$RPI = 100n\Sigma\Sigma EijDij/Nb$$

where n is the total number of patch types in a mosaic, Eij is the number of edges between patch types i and j, Dij is the dissimilarity value for patch types i and j, and Nb is the total number of edges of pixels (each pixel has four edges). This index measures the contrast of neighbouring patch types in a landscape mosaic. See Romme (1982) for an example of application.

Reciprocal Simpson's index of diversity ($1/S$) is the measure the probability of encountering two pixels with the same attributes when taking a random sample of two pixels:

$$1/S = 1/m\Sigma p_i^2$$

where pi is the fraction of area occupied by the attribute i, and m is the total number of attributes (Table 8.2).

Entropy (ENT) This index measures the disorder of pixels for each category:

$$ENT = -\Sigma\Sigma Pij*\ln(Pij)$$

where Pij is the probability of a grid point of land use i being adjacent to a grid point of land use j. Pij represents the probability that land use type i is adjacent to cells of type j. The Pij value is calculated by dividing the number of cells of type i that are adjacent to j by the total number of cells of type i present in the matrix:

$$Pij = Nij/Ni$$

where Nij is the number of adjacencies between pixels of patch type i and j. Ni is the total number of cells of type i. $Ni = \Sigma Nij$. Then $\Sigma Pij = 1$ (O'Neill *et al.* 1988, Turner *et al.* 1989, Li and Reynolds 1993).

The contagion index This index derives from the information theoretical measures (Shannon and Weaver 1962) and measures the degree of clumping; it also represents the deviation of the entropy measure from its possible maximum value:

$$C = 2 \ln m - ENT$$

where $ENT = -\Sigma\Sigma Pij*\ln(Pij)$ (see Entropy). This index measures whether cells are aggregated or clumped. $2 \ln m$ represents the maximum possible probability of adjacency (m = number of land cover categories). If the value of contagion is high it means that contiguous patches are found on the landscape. If the value is low the landscape is composed of small patches.

Relative contagion (RC) This index, proposed by Li and Reynolds (1993), utilizes the same components of the formula above but the entropy is divided by max entropy so that the RC varies between 0 and 1 and represents an evenness index:

$$RC = (1 - ENT)/2 \ln m$$

8.2.3 Distance measurements

The distance of a patch or a group of patches from others is an important parameter in mosaic analysis.

Table 8.2 Indices of spatial arrangement of the mosaic (from a subset of 8×8 km, see Fig. 8.2): $\Lambda(r)$ lacunarity $r = 2\times2$; pi = relative abundance

Land cover code	Abundance	pi	pi*log pi	$\Lambda(r)$	pi²
1	799	0.499	−0.347	1.52	0.249
2	34	0.021	−0.082	20	0.000
3	399	0.249	−0.346	2.34	0.062
4	125	0.078	−0.199	5.86	0.006
5	4	0.003	−0.015	138	0.000
6	16	0.010	−0.046	48	0.000
7	223	0.139	−0.275	4.08	0.019
Total	1600	1	−1.31		0.33

Data were processed using the routine in Basic of Boxes 8.2 and 8.3.
Diversity (Shannon diversity) = 1.31, Simpson diversity = 3.03, dominance = 0.464, contagion 2.28.

Distance means energy loss for moving, increased predation risk and decreased transportation by vectors etc. (see van Dorp and Opdam 1987). Distance also means connectedness and connectivity.

Distance can be calculated according to a combination of possibilities, as discussed in detail by Baker and Cai (1992) and summarized in Table 8.3. Distance can be measured according to a different selection of possibilities: from each patch to all the adjacent neighbours of each patch; from a patch to all others of the same group; from each patch to the single nearest patch of a different group; from a patch of a specific group to another patch of a specific group (Fig. 8.4 and Table 8.4).

Table 8.3 Measures utilized for calculating distances (gp = attribute group). (From Baker and Cai 1992, with permission)

Mean distance
Standard deviation distance
Mean distance by gp
Standard deviation distance by gp
Number of distances in each distance class
Number of distances in each distance class by gp

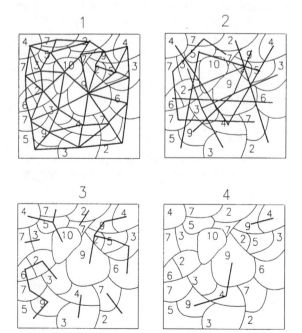

Figure 8.4 Four possible methods to measure distance (see text for explanation) (from Baker and Cai 1992, with permission).

Table 8.4 Measures of pixel attribute, patch size, shape, fractal dimension and perimeter (gp = attribute group). (From Baker and Cai 1992, with permission)

Attribute
    Mean pixel attribute
    Standard deviation pixel attribute
    Mean pitch attribute
    Standard deviation patch attribute
    Cover by gp
    Density by gp

Size
    Mean patch size
    Standard deviation size
    Mean patch size by gp
    Standard deviation size by gp
    Number in each size class
    Number in each size class by gp

Shape
    Indices
        Corrected perimeter/area
        Perimeter/area
        Related circumscribing circle
    Measures
        Mean patch shape
        Standard deviation shape
        Mean patch shape by gp
        Standard deviation shape by gp
        Number in each shape class
        Number in each shape class by gp

Fractal dimension
    Perimeter = area fractal dimension

Perimeter
    Sum of perimeters
    Sum of perimeters by gp
    Mean perimeter length
    Mean perimeter length by gp
    Standard deviation perimeter length
    Standard deviation perimeter length by gp

Proximity index (PX) The degree of isolation of patches is measured with the proximity index PX (Gustafson and Parker 1992):

$$PX = \Sigma(Sk/nk)$$

where Sk is the area of the patch and nk is the nearest neighbour distance of patch k. This index can be scaled as a proportion of the maximum value of PX, so that $PXs = PX/PX_{max}$, where $PX_{max} = E/n$, E is half of the total area of landscape separated by the minimum possible n. PX cannot be used to compare two landscapes of different extents because PX_{max} changes according to the landscape extent.

8.2.4 Texture measures

Texture measures are used to analyse patterns of brightness variation within an image (see Haralick *et al.* 1973, Musick and Grover 1991) (Table 8.5).

These measures can be used profitably in landscape ecology to analyse the complexity of the mosaic and the contrast between patches.

The spatial co-occurrence probability $p(i,j,d,\theta)$ is the probability that a pixel or a cell of type i is separated by a pixel or a cell of type j by distance d according to an angle direction θ, which may be 0° horizontal, 45° right diagonal, 90° vertical or 135° left diagonal. The comparison involves two reciprocal co-occurrences and the matrix produced is symmetrical. Figure 8.5 presents two examples of analysis of co-occurrences.

Two indices of homogeneity may be used in the analysis of landscape texture, the angular second moment (ASM) and the inverse difference moment (IDM). See Box 8.4 for routines in Basic to calculate *ASM* and *IDM* indices.

Angular second moment (*ASM*) The sum of co-occurrence probabilities:
$ASM = \Sigma\Sigma[p(i,j)]_2$ where $p(i,j)$ is the relative abundance of the cells i that are adjacent to cells j. $p(i,j) = n(i,j)/tot$, where $n(i,j)$ is the number of occurrences of cell i adjacent to cell j. $tot = \Sigma\Sigma n(i,j)$. See Box 8.2 for routine in Basic to calculate ASM.

ASM increases with mosaic homogeneity because the co-occurrence of identical values has a strong influence on this index. In homogeneous patches the co-occurrence of cells with identical values is dominant and the squared probabilities enhance this value. ASM has a value of 1 when all co-occurrences are identical, but this index is insensitive to the magnitude of the difference between cells of different value.

Inverse difference moment (*IDM*)
$IDM = \Sigma\Sigma[1/1 + (i-j)^2]\, p(i,j)$
This index measures the co-occurrences weighted according to the difference between values of i and j. The index has maximum value 1 when all cells or pixels are identical. To be usefully applicable in land analysis the differences in value of i and j must have some significance (intensity or interval type data). In the example in Fig. 8.2 the values of the landscape matrix from 1 to 7 have been ranked

Table 8.5 Value of co-occurence frequency matrix (A1, B1, C1) and the co-occurence matrix of probability A2, B2, C2 derived by the A1, B1, C1 matrix respectively dividing the frequencies by the total number of co-occurences. (From Musick and Grover 1991, modified, with permission)

A1

	21	22	23	24
21	4	2	0	0
22	2	0	0	1
23	0	0	0	1
24	0	1	1	0

B1

	21	22	23	24
21	10	7	1	1
22	7	2	1	3
23	1	1	0	1
24	1	3	1	0

C1

	21	22	23	24
21	6	8	0	2
22	8	2	2	2
23	0	2	0	2
24	2	2	2	0

A2

	21	22	23	24
21	0.333	1.67	0	0
22	0.167	0	0	0.083
23	0	0	0	0.083
24	0	0.083	0.083	0

B2

	21	22	23	24
21	0.250	0.175	0.025	0.025
22	0.175	0.050	0.025	0.075
23	0.025	0.025	0	0.025
24	0.025	0.075	0.025	0

C2

	21	22	23	24
21	0.150	0.200	0	0.050
22	0.200	0.050	0.050	0.050
23	0	0.050	0	0.050
24	0.050	0.050	0.050	0

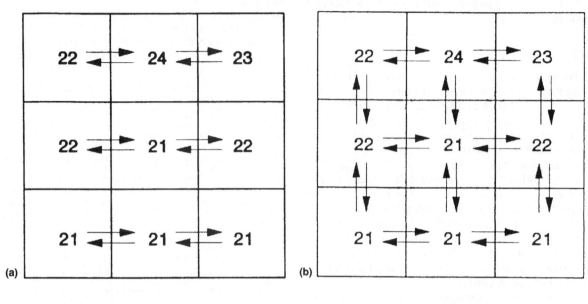

(a)

(b)

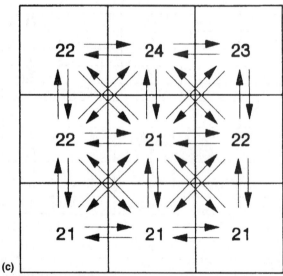

(c)

Figure 8.5 Three possibilities to calculate the co-occurrence probability $p(i,j,d, \theta)$ between cells or pixels of a matrix. The number in the cells indicates the type of attribute, which may be land cover, vegetation, or colour. In (**a**) the co-occurrences have been measured only along the horizontal axis ($\theta = 0°$); in (**b**) according to the four perpendicular directions ($\theta = 0, 90°$) and in (**c**) in all directions ($\theta = 0, 45, 90, 135°$) at a distance $d = 1$ (from Musick and Grover 1991, modified, with permission).

according to biomass cover (1 dense beech forest to 7 rock outcrops).

Contrast (*CON*)

This index measures the contrast present in the landscape:

$$CON = \Sigma\Sigma[(i-j)^2 * Pij]$$

See Box 8.4 for routine in Basic to calculate *CON*.

Change index (*C*)

This measures the change of surface of a patch over time:

1,1,1,1,1,1,1,1,1,1,1,1,1,4,1,1,1,1,1,1,1,4,1,1,1,1,1,1,1,1,4,1,3,1,1,1,1,1,1
1,1,1,1,1,1,1,1,1,1,1,1,1,4,4,1,1,1,1,1,1,1,1,1,1,1,1,1,1,1,1,1,4,3,1,1,1,1,1
1,1,1,1,1,1,1,1,1,1,1,1,1,4,4,4,4,1,1,1,1,1,1,1,1,1,1,1,3,3,3,4,1,1,1,3,3,1,1,1,1
1,1,1,1,1,1,1,1,1,3,3,3,3,3,3,4,1,4,4,4,1,1,1,1,3,3,4,4,4,2,1,1,1,1,1,1,1,3,3,1,3,1
3,1,1,1,1,1,1,1,1,1,3,3,3,3,1,1,1,1,1,1,1,1,1,1,1,3,3,1,1,3,3,1,1,1,1,1,4,2,1,4,1,1,1
7,3,1,1,1,4,4,1,1,1,4,3,1,1,1,1,1,1,1,1,1,1,1,3,3,2,3,1,3,3,3,3,1,1,4,4,1,1,1,1,1
7,7,7,7,7,2,3,1,1,1,1,1,1,1,1,6,1,1,3,1,1,1,1,3,3,3,1,1,1,3,7,3,1,1,1,1,1,1,1,1,1
7,7,7,7,1,3,7,1,1,1,1,1,1,1,1,6,1,1,3,1,1,1,1,3,1,3,1,3,1,1,1,7,1,1,1,1,1,1,1,1,1
1,7,3,1,3,1,1,1,1,1,1,1,1,1,1,1,1,4,3,1,1,1,1,1,1,1,1,1,1,1,3,1,1,1,1,4,4,4,4,1
1,7,1,1,1,1,1,1,1,1,1,1,1,1,1,1,1,4,1,6,3,3,1,3,3,3,3,4,4,1,1,3,1,1,1,1,1,1,1,4,4
3,7,1,1,1,1,1,1,1,1,1,1,1,1,1,4,1,1,1,3,4,1,1,3,4,4,1,1,1,3,1,1,1,1,4,4,4,1,4,4,1
7,1,1,4,4,4,1,4,4,1,1,1,1,1,1,7,1,1,1,1,1,4,1,1,1,4,1,1,1,1,1,1,1,1,1,1,4,4,4,1,4,1
3,1,1,1,1,4,4,4,1,1,1,1,1,1,7,1,1,1,1,1,4,1,1,1,4,1,1,1,1,1,1,1,1,1,7,7,7,1,1,1,1,1
1,1,1,3,7,7,3,3,1,1,7,3,7,7,7,7,3,3,1,1,1,1,1,1,1,1,1,1,1,1,1,1,1,3,3,7,4,1,1,1,1,4,1
1,1,7,7,7,7,3,3,3,4,7,7,7,1,3,3,3,3,3,1,1,1,1,1,1,1,1,1,1,1,1,3,3,7,7,3,3,7,3,3,1
7,7,7,7,7,7,3,2,7,7,7,7,3,4,3,3,1,3,1,3,1,1,1,1,1,1,1,1,1,1,1,1,1,3,3,1,3,3,3,3,3,3
1,4,7,7,3,7,7,3,3,1,4,3,3,2,3,3,7,1,3,7,7,7,1,1,1,1,1,1,3,1,1,1,1,1,1,1,2,4,3,1,1
1,7,7,7,7,7,7,7,7,1,3,3,3,3,3,1,3,1,3,3,7,3,1,1,1,1,1,1,1,3,3,1,2,2,1,1,1,4,1,1,1
1,1,7,7,7,7,7,7,1,3,7,3,7,7,1,1,1,1,1,1,1,1,1,1,1,1,1,1,3,3,2,2,2,1,1,1,4,1,1,1
7,7,7,1,1,3,7,7,1,1,7,7,7,3,7,3,3,1,1,7,3,1,1,1,1,1,1,1,1,1,1,1,3,2,2,2,1,3,3,1,4,6,1,1
7,7,3,1,3,3,7,1,1,1,3,3,1,1,7,3,3,7,7,1,7,1,1,1,1,1,1,1,1,1,2,2,2,3,2,3,1,6,6,6,6,1
3,1,4,1,1,1,3,1,3,3,3,3,1,7,7,3,3,7,3,3,3,4,1,1,1,1,1,1,3,3,3,3,3,3,1,1,3,3,1,1,1,6,6
2,1,4,4,1,1,3,3,7,7,3,3,3,3,1,1,3,7,3,1,3,7,4,1,1,1,1,3,3,3,3,4,1,3,3,3,7,3,1,1,1,6
3,3,3,4,1,3,3,3,3,3,3,3,4,7,1,1,1,7,1,4,7,1,1,3,1,1,3,3,3,3,3,1,3,1,4,1,1,1,1,1,1,1
3,3,3,1,3,3,3,3,3,3,3,1,3,7,3,7,3,1,1,1,4,4,7,7,3,1,1,3,3,1,7,7,7,7,7,7,3,1,1,1,1,4
3,1,3,3,3,3,3,3,3,3,3,1,3,4,3,3,3,7,1,1,7,3,1,2,2,7,7,1,1,3,7,7,7,7,7,7,7,7,3,1,4
2,3,3,3,3,3,1,7,3,3,5,4,1,4,1,7,7,3,3,7,7,3,3,7,7,7,7,1,1,1,3,7,3,3,3,7,7,7,1
3,3,3,3,3,1,1,3,3,1,1,4,4,1,7,7,7,7,7,3,3,3,1,1,3,1,1,3,1,1,1,1,3,1,7,7,3,3,3,3
3,3,3,3,1,1,1,3,3,4,4,4,2,7,7,7,1,7,3,3,3,1,1,1,1,4,1,1,1,1,1,3,3,3,3,1,3,1,3
3,3,3,3,1,1,3,1,3,4,4,7,7,2,1,3,3,3,1,1,7,3,3,4,1,1,1,1,1,1,1,3,7,7,1,1,1,7
1,1,4,3,3,3,3,3,1,1,1,1,7,2,7,7,3,3,1,3,7,1,3,2,3,1,1,1,1,1,6,6,6,1,7,1,1,1,3,7
1,1,1,4,5,3,1,1,1,1,1,4,4,7,7,3,3,1,1,7,4,7,3,1,1,3,1,4,1,1,1,1,6,1,1,1,1,3,1,3
7,1,1,1,1,3,7,7,3,1,1,7,7,7,3,3,3,1,1,7,7,1,1,1,3,7,1,3,3,1,1,1,1,1,6,1,1,1,3,3
1,3,3,4,4,1,7,3,3,3,7,7,3,3,3,3,3,4,7,3,1,1,1,3,1,3,1,3,1,1,1,1,1,1,1,1,7,7,7
3,7,7,7,1,1,1,3,3,7,7,3,3,3,3,3,3,7,1,4,1,4,7,3,1,1,1,1,2,1,3,1,3,1,1,1,1,1,1,3,4
1,7,7,4,1,1,1,3,3,2,3,3,3,1,1,7,7,1,7,7,3,1,3,1,1,7,7,7,1,1,1,1,3,1,1,1,1,3,1,1,1,1
1,4,4,4,1,1,1,3,2,3,3,1,1,1,3,3,7,7,7,1,7,7,7,1,3,7,7,1,1,1,1,7,7,3,1,1,1,1,1,1,3
1,1,1,3,1,1,1,3,2,3,3,1,1,3,3,3,3,3,7,7,4,4,7,7,7,3,3,1,1,1,3,7,7,3,3,1,1,1,1,7
7,3,3,3,1,1,1,1,2,1,1,1,1,1,1,3,3,2,2,3,3,4,3,3,7,7,1,1,1,1,3,3,7,7,1,1,1,1,4,3
7,3,3,4,1,4,3,1,1,5,4,4,1,3,3,3,3,3,3,4,1,6,3,1,3,3,3,1,1,1,1,3,3,7,1,1,3,3,1,1,5

Figure 8.6 Numerical map (40 × 40 cells) as transformed from the map MacGIS of Fig. 8.2. The numerical values can be imported into Basic routines for further processing, e.g. to calculate richness (S), H′ diversity, inverse Simpson diversity (SP), entropy (ENT), contagion (C), angular second moment (ASM), inverse difference moment (IDM), and contrast (CON). Code numbers: 1 = oak woodland, 2 = beech woodland, 3 = cultivations, 4 = prairies, 5 = urban, 6 = rock outcrops, 7 = mixed cultivation.

$$C = ((pk2 - pk1)/(t2 - t1))/n$$

where $pk2$ is the surface of category k in time 2 and $pk1$ the same category in time 1; $t2$ and $t1$ are respectively the dates of the time lag.

Figures 8.6, 8.7, 8.8 and Tables 8.6 and 8.7 describe an example of the application of spatial indices to the study of landscape mosaic.

8.2.5 Semivariance

Semivariograms are used to measure variance at many scales, comparing the values of a random variable at two points at a given lag distance. Semivariograms are used mostly in geostatistics (Isaaks and Srivastava 1989):

$$g(\gamma) = 1/2N(\gamma)\Sigma(Xj-(Xj+\gamma)^2)$$

where $g(\gamma)$ is the semivariance at lag γ, $N(\gamma)$ is the number of pairwise comparisons at lag γ, and Xj is the random variate at position j.

The plot of semivariance g against lag γ allows us to see at which distance the variance changes. The semivariance generally increases with increasing distance, although this is not true for all processes and is inversely related to the spatial autocorrelation of a variable.

From the two pictures in Figure 8.9 it is possible to assess that at a distance of about 150 km there is a maximum difference in density and that this decreases when we move to a biogeographic scale. Data from the Breeding Bird Survey (BBS) have no asymptotic maximum semivariance and the maximum is reached at an intermediate distance. This can mean that a species is more abundant around the central point of biogeographic distribution, and that away from this point the abundance decreases, showing a higher spatial autocorrelation (Villard and Maurer 1996).

8.2.6 Lacunarity to measure landscape texture

Deterministic fractals with identical dimensions can have a different appearance, as in the case of Cantor dusts with a difference of 0.5.

Mandelbrot called the distribution of gap size lacunarity (Mandelbrot 1982). Lacunarity, then, measures the distribution of gaps in a fractal figure. An object with a low lacunarity is invariant when translating; on the other hand, an object with a heterogeneous gap size is not translationally invariant. But we must consider that the invariance is scale dependent: an object invariant at a small scale may be heterogeneous at broad scale and vice versa. Translational invariance is not synonymous with self-similarity.

Lacunarity has advantages compared to other indices of landscape structure (Plotnick et al. 1993):

● The algorithm is relatively simple.
● The gliding box algorithm samples the map sufficiently to quantify change in contagion and self-similarity with scale.

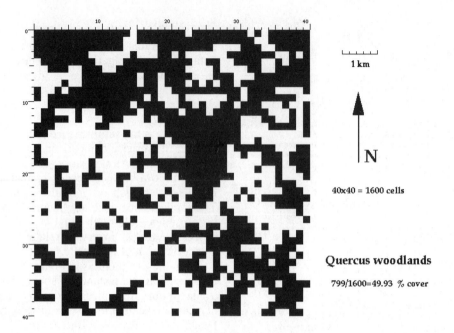

Figure 8.7 Example of landscape analysis using a single land cover, in this case oak woodland (source from Nardelli 1995). See also Fig. 8.2 for the complete mosaic. From this image it is possible to measure: number of patches, area, perimeter, Shannon diversity of patch size, Shannon diversity of patch edge, max, min and mean patch size and patch edge. See Tables 8.7 and 8.8.

- The results are not sensitive to boundary maps.
- The analysis can be used for very sparse data.

The fragmentation that is one of the major human-induced disturbance effects can change the

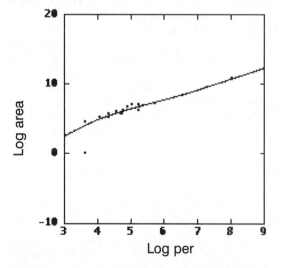

Figure 8.8 Plot of log area (natural logarithm of patch area) against log per (natural logarithm of patch perimeter). Smoothed using Systat DWLS method. Data from Fig. 8.7.

heterogeneity of a landscape, causing species distribution. Lacunarity seems a reasonable method to measure this heterogeneity.

Cantor dust is a fractal object obtained by applying a generator to a unit interval that divides into three parts of equal intervals. The middle part is deleted. The two parts are then divided again, each into three parts. In five generations the segment lengths are so short that it is not possible to distinguish them in the sixth generation. The fractal dimension of the triadic Cantor set is a fractal set with a fractal dimension $D = \ln2/\ln3$.

To calculate lacunarity we utilize the gliding box algorithm according to Allain and Cloitre (1991) (see also Plotnick *et al.* 1993).

A box $r \times r$ is moved from the upper left hand corner to the bottom right hand corner by steps of a cell for each column, and the number of occupied sites calculated according to classes of box mass (number of occupied sites):

$$\Lambda(r) = Z^{(2)}/(Z^{(1)})^2$$
$$Z^{(1)} = \Sigma SQ(S,r)$$
$$Z^{(2)} = \Sigma S^2 Q(S,r)$$

The probability distribution is:
$$Q(S,r) = n(S,r)/N(r)$$

Table 8.6 *Quercus* woodland cover, analysis of patch size and perimeter (see Fig. 8.7 for more details).
No patch = code of individual patch;
Area = surface of patches;
Perimeter = length of patch edge;
Log Area = natural logarithm of patch size
Log Perimeter = natural logarithm of patch edge;
PiArea = relative importance of each patch:
PiPerimeter = relative importance of each patch edge;
piArea*log(piArea) = to find patch size diversity (Shannon diversity);
piPerimeter*log(piPerimeter) = to find patch edge diversity (Shannon diversity);
H'a = Shannon diversity of patch size – ΣpiArea*log(piArea);
H'p = Shannon diversity of patch edge – ΣpiPerimeter*log(piPerimeter).

no patch	Area	Perimeter	Log Area	Log Perimeter	PiArea	PiPerimeter	piArea* log(piArea)	piPerimeter* log(piPerimeter)
1	100	37.660	4.605	3.629	0.001	0.005	−0.008	−0.026
2	100	37.660	4.605	3.629	0.001	0.005	−0.008	−0.026
3	100	37.660	4.605	3.629	0.001	0.005	−0.008	−0.026
4	100	37.660	4.605	3.629	0.001	0.005	−0.008	−0.026
5	100	37.660	4.605	3.629	0.001	0.005	−0.008	−0.026
6	100	37.660	4.605	3.629	0.001	0.005	−0.008	−0.026
7	100	37.660	4.605	3.629	0.001	0.005	−0.008	−0.026
8	100	37.660	4.605	3.629	0.001	0.005	−0.008	−0.026
9	200	57.660	5.298	4.055	0.003	0.007	−0.015	−0.036
10	200	57.660	5.298	4.055	0.003	0.007	−0.015	−0.036
11	200	57.660	5.298	4.055	0.003	0.007	−0.015	−0.036
12	200	75.310	5.298	4.322	0.003	0.010	−0.015	−0.045
13	290	110.970	5.670	4.709	0.004	0.14	−0.021	−0.061
14	300	76.490	5.704	4.337	0.004	0.010	−0.021	−0.045
15	300	112.970	5.704	4.727	0.004	0.015	−0.021	−0.062
16	400	96.490	5.991	4.569	0.005	0.012	−0.027	−0.054
17	500	117.660	6.215	4.768	0.006	0.015	−0.032	−0.063
18	500	188.280	6.215	5.238	0.006	0.024	−0.032	−0.090
19	500	116.490	6.215	4.758	0.006	0.015	−0.032	−0.063
20	780	189.800	6.659	5.246	0.010	0.024	−0.046	−0.091
21	800	134.140	6.685	4.899	0.010	0.017	−0.046	−0.070
22	1000	210.630	6.908	5.350	0.013	0.027	−0.055	−0.098
23	1100	154.410	7.003	5.038	0.014	0.020	−0.059	−0.078
24	1141	187.800	7.040	5.235	0.014	0.024	−0.061	−0.090
25	1300	304.770	7.170	5.720	0.016	0.039	−0.068	−0.127
26	4350	706.710	8.378	6.561	0.055	0.091	−0.160	−0.218
27	14591	1443.710	9.588	7.275	0.185	0.186	0.312	−0.313
28	49602	3071.670	10.812	8.030	0.627	0.395	−0.292	−0.367
Total	79054	7772			1	1	H'a 1.41	H'p 2.25

where $n(S,r)$ is the number of boxes in which a box mass category has been found.

The frequency distribution is:

$$N(r) = (M - r + 1)2$$

where the number of boxes of size r containing S occupied sites is indicated by $N(r)$. M = size of the map.

The lacunarity index is better understood by considering it as a ratio of $\Lambda(r) = Z^{(2)}/(Z^{(1)})^2$, where $Z^{(1)} = S(r)$ is the mean box mass and $Z^{(2)} =$

Table 8.7 Minimum (min), maximum (max) and mean of patch size area and patch edge (perimeter) of oak woodland cover (see Table 8.6, Fig. 8.2 and Table 8.2 for more details)

	Patch size (area)	Patch edge perimeter
min	100	37.660
max	4962	3071.67
mean	2823	277

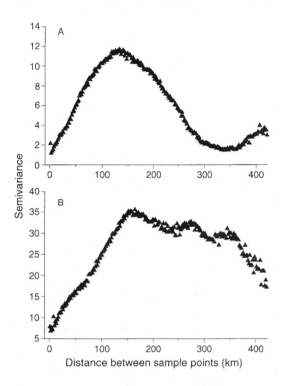

Figure 8.9 The semivariance has been calculated for two species of birds: red-headed woodpecker (**A**) and red-bellied woodpecker (**B**) (from Villard and Maurer 1996, with permission).

$s_s^2(r) + S^2(r)$, $S(r)$ is the mean and $s_s^2(r)$ the variance of the number of sites per box. Then, $\Lambda(r) = s_s^2(r)/S^2(r) + 1$.

1. Lacunarity is a function of the gliding box. A larger box size has a lower lacunarity because by increasing the size of the box the relative variance decreases.
2. Lacunarity is a function of occupied sites. The lower the fraction P of occupied sites the higher the lacunarity $s_s^2(r)/S^2(r)$ goes to infinity.
3. Lacunarity changes according to the spatial arrangement of occupied sites. Large patches have higher lacunarity values than sparse patches.

Lacunarity analysis may be made using different maps with the same r, or different gliding box sizes for the same map.

Simulations conducted by Plotnick *et al.* (1993) on simulated maps with r ranging from 1 to 128 have demonstrated that the highest value of lacunarity is observed when $r = 1$ and the grain size of the maps is equal to the box size.

Lacunarity can also be used to analyse transect data. In this case the gliding box is substituted by a linear lag (Fig. 8.10).

8.2.7 Measuring boundaries in the landscape

In an earlier chapter we considered ecotones as important structures in a landscape. These are inherent properties in a landscape but also function as shaping factors in many processes.

The detection of boundaries in a landscape is not a simple matter. In fact, the edges between two different habitats or types of land cover never have the functions of true boundaries. On the other hand, especially in human-dominated landscapes, the boundaries are so narrow and the habitat constraints so high that is difficult to find a correlation between boundary structure and function.

It appears clear from the study of animal behaviour that often a boundary is perceived in the neighbourhood of a physical edge. So, the distance from this environmental discontinuity can be really important in the life history of a species.

The complexity of a landscape can be measured in terms of patch diversity and also in terms of boundary complexity. Owing to the recognized importance of boundaries in the dynamics and functioning of a landscape, their measurement is a fundamental step towards acquiring a knowledge of the structure and functioning of the land mosaic. Recently Metzger and Muller (1996) elaborated a method for measuring certain relevant characters of a landscape, assessed by remote sensing technology and presenting several indices of boundary proportions, land cover boundary complexity and landscape boundary complexity. More details are given in Section 8.5.3.

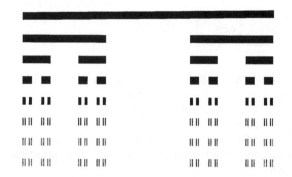

Figure 8.10 Cantor dust. This model has a fractal dimension equal to $\ln N/\ln(1/r)$, where N is the number of small pieces generated and r is the size of reduction at each iteration.

Box 8.1

PROGRAM TO CALCULATE PATCH CHARACTERS
This routine has been prepared in Basic

```
REM MPC MEASURE OF PATCH CHARACTERISTICS
REM Program to calculate:
REM L/S ratio
REM Corrected Perimeter/Area CPA=(.282*L)/√S
REM Related circumscribing circle RCC=2*(area/π)½/longest axis
REM Shape Index S1 1/Ni*Σ(Li/Si)
REM Shape Index S2 = 1/Ni*(Li/4√Si)
REM data input
REM S=Patch area
REM Si=Patch area of the category i
REM L=Perimeter
REM Ni=number of patches of category i in a map
REM Ax=Longest Axis
REM that.file is the name of your data file
REM Number of patches

OPEN "I", #1, "that.file" 'your file data
100 if eof(1) then end: goto 1000
cc=cc+1
INPUT #1, area, perimeter, laxis

a(cc)=area: p(cc)=perimeter: ax(cc)=laxis

goto 100
1000
FOR j=1 TO cc
ls=p(j)/a(j)
CPA=(.282*p(j))/SQR(a(j))
RCC=2*SQR((a(j)/3.14))/ax(j)
tot=tot+ls
p(j)=p(j)/4
a(j)=SQR(a(j))
tot1=tot1+p(j)/a(j)
PRINT USING "##.##"; "First shape index: C/S"; ls
PRINT USING "##.##"; "Second shape index: CPA"; corr
PRINT USING "##.##"; "Third shape index: CPA"; RCC
NEXT j
S1=1/(cc*/tot): 'Fourth shape index
S2=1/(cc*tot1): 'Fifth shape index
PRINT USING "##.##"; "Fourth shape index: S1"; S1
PRINT USING "##.##"; "Fifth shape index: S2"; S2

END
```

Box 8.2

PROGRAM TO CALCULATE NUMBER AND RELATIVE IMPORTANCE OF COVER TYPE, SHANNON DIVERSITY, DOMINANCE INDEX AND INVERSE SIMPSON DIVERSITY

This routine has been prepared in Basic

```
REM diversity routine
REM This routine calculates:

REM 1. Number of cover types: land cover, vegetation, others
REM 2. Relative abundance of cover types
REM 3. Shannon diversity
REM 4. Dominance index
REM 5. Inverse Simpson diversity

REM input file, as sequential file produced by exportation from MacGis
REM using the routine export in text format
REM then with Word the space between values are filled with a coma
REM the file is saved as text and it is ready to be read by this routine

DIM cod1(100), cod2(100

OPEN "I", #1, "that.file"

100 IF EOF(1) THEN GOTO 1000

INPUT #1, A
cod1(A)=cod1(A)+1

GOTO 100

1000 'number of land uses

FOR I=1 TO 100
IF cod1(I)><0 THEN tot=tot+1
NEXT I
PRINT "Number of cover type categories"; tot

FOR I=1 TO tot
total=total+cod1(I)
REM total = number of land use categories
NEXT I
REM tot = number of patches belonging to all land uses

FOR I=1 TO tot
cod2(I)=cod1(I)/total
cod3(I)=cod2(I)*LOG(cod2(I))
```

```
PRINT I,: PRINT USING "#.###";cod2(I)

NEXT I

FOR I=1 TO tot
h=h+cod3(I)
simpson=simpson+cod2(I)*cod2(I)
NEXT I

PRINT "Shannon diversity   ",: PRINT USING "#.###";-h
PRINT "Dominance   ",: PRINT USING "#.###";LOG(tot)-ABS(h)
PRINT "Inverse Simpson   ",: PRINT USING "#.###";1/simpson
```

Box 8.3

PROGRAM TO CALCULATE CONTAGION INDEX
This routine has been prepared in Basic

```
REM CONTAGION ROUTINE
Input "Number of rows"; YY
Input "Number of columns"; xx
Input "Number of land uses"; nlu
DIM cod(yy+1,xx+1), contagion(nlu,nlu), pi(nlu,nlu)
l=1
OPEN "I", #1, "that.file

100 IF EOF(1) THEN GOTO 1000
INPUT #1, a
cod1(a)=cod1(a)+1

k=k+1
cod(l,k)=cod(l,k)+a

IF k=XX THEN k=0:l=l+1

GOTO 100

1000 '
REM j = code of land use category
FOR j=1 TO nlu
PRINT j
```

```
FOR k=1 TO YY
FOR l=1 TO XX

IF cod(l,k)=j THEN GOSUB 3000

NEXT l
NEXT k

NEXT j

FOR i=1 TO nlu
FOR l=1 TO nlu
'
pi(i,l)=pi(i,l)+contagion(i,l)/cod1(i)
IF pi(i,l)><0 THEN ENT=ENT+pi(i,l)*LOG(pi(i,l))

NEXT l

NEXT i
PRINT "Contagion",:PRINT USING "##.###"; 2*LOG(nlu)+ENT

END

3000'
IF cod(l,k+1)><cod(l,k) THEN contagion(j,cod(l,k+1))=contagion(j,cod(l,k+1))+1
IF cod(l,k-1)><cod(l,k) THEN contagion(j,cod(l,k-1))=contagion(j,cod(l,k-1))+1
IF cod(l+1,k)><cod(l,k) THEN contagion(j,cod(l+1,k))=contagion(j,cod(l+1,k))+1
IF cod(l-1,k)><cod(l,k) THEN contagion(j,cod(l-1,k))=contagion(j,cod(l-1,k-1))+1:
RETURN
```

Box 8.4

PROGRAM TO CALCULATE ASM (ANGULAR SECOND MOMENT), IDM (INVERSE DIFFERENCE MOMENT) AND CON (CONTRAST)
This routine has been prepared in Basic

```
REM This routine finds the value
REM of ASM Angular Second Moment, IDM Inverse Difference Moment and CON Contrast

INPUT "col"; col: 'Column
INPUT "row"; row: 'Row
INPUT "land cover type"; lct
REM The cell adjacency has ben calculated with θ=0 along the horizontal axis (see Fig. 8.5)

DIM cod(row+1,col+1), p(lct+1,lct+1), pi(lct+1,lct+1)
```

```
l=1
OPEN "I", #1, "that.file

100 IF EOF(1) THEN GOTO 1000
INPUT #1, a

k=k+1
cod(l,k)=cod(l,k)+a

IF k=col THEN k=0:l=l+1

GOTO 100

1000'
FOR k=1 TO col
FOR l=1 To row

p(cod(l,k),cod(l,k+1))=p(cod(l,k),cod(l,k+1))+1
p(cod(l,k),cod(l,k–1))=p(cod(l,k),cod(l,k–1))+1

p1(cod(l,k),cod(l,k+1))=p(cod(l,k)–cod(l,k+1))
p1(cod(l,k),cod(l,k–1))=p(cod(l,k)–cod(l,k–1))

NEXT l
NEXT k

FOR i=1 TO lct
FOR l=1 TO lct
cod1=cod1+p(i,l)

NEXT l
NEXT i

FOR i=1 To lct
FOR l=1 TO lct

pi(i,l)=pi(i,l)+(p(i,l)/cod1)*(p(i,l)/cod1)

asm=asm+pi(i,l)
p2(i,l)=p1(i,l)^2

IDM=IDM+p(i,l)/cod1*(1/1+p2(i,l)))
CON=CON+p(i,l)/cod1*p2(i,l)

NEXT l
NEXT i
PRINT "ASM",:PRINT USING "##.##";asm
PRINT "IDM",:PRINT USING "##.##";IDM
PRINT "CON",:PRINT USING "##.##";CON

END
```

Box 8.5

PROGRAM TO CALCULATE LACUNARITY INDEX
This routine has been prepared in Basic language

```
REM LACUNARITY ROUTINE
REM Input information on land cover (Code), Box size (R) and Matrix size (W)
REM W=Number columns×number of rows
INPUT "code of land use or vegetation"; code
INPUT "Box size";R
INPUT "Matrix size";W

DIM cod(W,W), n(R^2), n2(R^2), n3(R^2)

l=1

REM prova.txt is a sequential file in which data are in the format:
REM a=1
REM a=1
REM a=2
REM a=3
REM....
OPEN "I", #1, "that.file

100 IF EOF(1) THEN GOTO 1000

INPUT #1, a: 'Reading data in the file

IF a><code THEN a=0: 'selection of land cover type
IF a=code THEN a=1
k=k+1
cod(l,k)=cod(l,k)+a
IF k=W THEN k=0:l=l+1
GOTO 100

1000'
nr=(W-R+1)^2
FOR l=1 TO W-(R-1)
FOR k=1 TO W-(R-1)

FOR i=0 TO R-1
FOR j=0 TO R-1

TOT=TOT+cod(l+i,k+j)
```

```
NEXT j
NEXT i
box=R*R

FOR m=0 To box
IF TOT=m THEN n(m)=n(m)+1
NEXT m
TOT=0

NEXT k
NEXT l

FOR m=0 TO box
n1(m)=n(m)/nr
IF m>0 THEN n2(m)=m*n(m)/nr
If m>0 THEN n3(m)=m^2*n(m)/nr
Z1=Z1+n2(m): Z2=Z2+n3(m)
NEXT m

lacunarity=Z2/Z1^2: PRINT "LACUNARITY"; lacunarity
```

8.3 THE FRACTAL GEOMETRY APPROACH

8.3.1 Introduction

Heterogeneity is a common pattern of the environment and is particularly visible at the landscape scale. Organisms, populations and communities have spatial distributions that reflect the heterogeneous nature of the land.

The unequal distribution of natural phenomena, such as the geological nature of rocks, the rain distribution across a mountainous range or the distribution of tree cover in a watershed, creates complicated mosaics to which organisms react. To measure this complexity Euclidean geometry often seems inadequate and new approaches are required; in this fractal geometry seems to fit the bill (Mandelbrot 1982, Feder 1988, Milne 1991, Hastings and Sugihara 1993).

In a manmade landscape in which straight lines and regular geometric figures have been created, transforming wild land into rural or urbanized areas, Euclidean geometry may be used to describe simple spatial patterns such as perimeter/area ratio, patch area and patch distance. When we consider a more natural landscape such figures disappear, and the irregularity of patches reduces most of the descriptive capacities of Euclidean geometry.

Fractal geometry brings a new perspective to the study and interpretation of landscape complexity and dynamics across scales. The aim of this chapter is to introduce the use of fractal geometry into landscape research, presenting a simplified view of a very complicated mathematical approach and reporting examples from a large variety of scales, from landscape to individuals.

Fractal geometry is so useful in landscape analysis because the hierarchical complexity of landscapes and their scaled patterns and processes need powerful tools to be investigated. Fractals are represented not only by patterns such as forest patch shape, but also by processes, and this last component is extremely useful. Fractal analysis can be applied to patch shape and spatial arrangement, but also to the distribution of animals in the space.

8.3.2 Concepts and definitions

The word fractal was coined in 1975 by Mandelbrot to describe an irregular object in which the irregularity is present at all scales, i.e. scale invariant. Mandelbrot (1986) proposed this definition: 'A fractal is by definition a set for which the Hausdorff Besicovitch dimension strictly exceeds the topological dimension'. A fractal is a shape made of parts similar to the whole in some way. Fractals can be considered objects or patterns that have non-integer dimensions. When a fractal object has qualities of the patterns at coarse scale that are repeated at finer and finer scale, the object shows a *self-similarity*.

Two different types of fractals can be observed: regular and random. The first type is represented by scale-invariant (self-similarity or self-affinity) objects. Regular fractals have exact self-similarity. When an object is a rescaled copy of itself in all directions (isotropic) it has a self-similarity attribute. When the rescaling is anisotropic the object presents a self-affinity.

The second category pertains to natural fractals (clouds, coastlines, organism abundance in the space etc.). Generally most natural fractals deviate from linear self-similarity and are called random fractals, displaying a statistical version of self-similarity.

Related to self-similarity is the concept of scale dependence. For instance, the coast is a fractal object for which the total length depends on the scale of resolution at which the measure is taken. The complexity is measured with the fractal dimension D, which is the counterpart of the familiar Euclidean dimensions 0 (point), 1 (line and curves), 2 (surfaces), 3 (volumes), and it is never integer (Figs 8.11 and 8.12). In regular one-dimensional objects the mass increases in proportion to the length, say $2R$. The mass in a two-dimensional disc with radius R increases in proportion to πR^2, the area of a circle; in a three-dimensional object the mass increases by $4/3\pi R^3$, that is, the volume of a sphere. Adding dimensions, the mass increases according to the power of the number of dimensions. In a fractal object R is raised to some power Dm that is not an integer number.

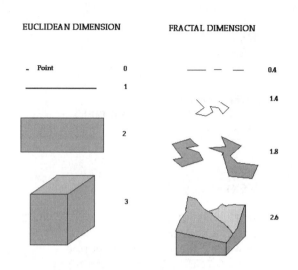

Figure 8.11 Comparison between Euclidean dimension and fractal dimension.

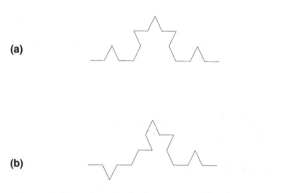

Figure 8.12 Example of regular (**a**) and randomized (**b**) Koch snowflakes.

The fractal approach is intuitively easy to understand but it is necessary to develop and apply the theory to practice. For further information on fractal geometry we recommend Mandelbrot (1982), Hasting and Sugihara (1993), Feder (1988) and Frontier (1987).

Fractal geometry finds a broad range of applications in different disciplines of the natural sciences, such as geology (Acuna and Yortsos 1995, Loehle and Li 1996), hydraulics (Ichoku *et al.* 1996), soil composition (Perrier *et al.* 1995, Barak *et al.* 1996), dynamics (Perfect and Kay 1995, Rasiah 1995, Perfect *et al.* 1996) microbial transport (Li *et al.* 1996) and vegetation structure (Chen *et al.* 1994). It is particularly useful in the study of phenomena that are ambiguous according to their scalar properties. Coastline length is a classic example. The length of the coast depends on the measuring unit. In this case, increasing the size of measuring unit, e.g. from metre to kilometre, the total length of the coast decreases. So, the length of a coast is scale dependent and relating this measure to the size of an organism, such as a sea otter or a crab, is possible by adopting the correct measure scale to which an organism is sensitive. This consideration can also be made for human-related phenomena: for instance, the number of suitable harbours along the coast decreases as the size of the ship increases.

Many patterns and processes are scale dependent and fractal models can describe their characteristics without the ambiguity of Euclidean geometry. For this reason, fractals seem more and more important in landscape ecology and related sciences.

The scale properties of objects measured using fractal geometry need more clarification.

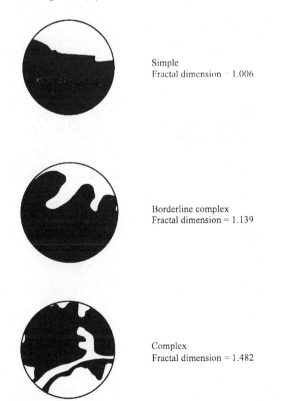

Simple
Fractal dimension = 1.006

Borderline complex
Fractal dimension = 1.139

Complex
Fractal dimension = 1.482

Figure 8.13 Examples of different complexity of a vegetation border expressed by the fractal dimension *D*; note that the increase in edges is equivalent to the increase in fractal dimension (from van Hees 1994, with permission).

Components of scale are lag, window, spatial and/or temporal extension of observed quantities and the grain of resolution (Turner *et al.* 1991). Other details are presented in Chapter 3.

Fractal models can be applied to measure landscape characters, but also to measure patterns perceived as species specific (Johnson *et al.* 1992). Both approaches are extremely useful in understanding the complexity of the environment and to predict species-specific responses to spatial configurations of resources (Fig. 8.13).

Examples of application to riparian forest patches are shown by Rex and Malanson (1990); Leduc *et al.* (1994) combined fractal analysis with variogram techniques to estimate the fractal dimension of a fragmented landscape; van Hees (1994) measured the complexity of Alaskan vegetation by applying the fractal technique of the dividers method. Lathrop and Peterson (1992) used the fractal approach to identify structural self-similarity in a mountainous landscape, measuring the area/perimeter relationship.

In the next section examples of the application of fractal geometry to landscape structures and to animals moving and acting in the landscape will be discussed.

8.3.3 The fractal dimension of the edges

Many processes and organisms are sensitive to patch shape but patch convolution is difficult to measure using Euclidean geometry. The fractal approach to the study of edge complexity takes into account the scale at which an edge is measured and the length of the ruler we use to measure. In other words, the length of objects such as coastlines, rivers and mountain ridges depends on measurement scale *L*. According to a simple power law:

$$C(L) = kL^{1-D}$$

where $C(L)$ is the length of the object (coastline, river etc.) measured at scale *L*, $D(2 > D > 1)$ is the fractal dimension and *k* is a constant. Increasing *L* reduces the total length $C(L)$, and vice versa if *L* is small.

For a Euclidean object $D = 1$, the length is independent of measurement scale. For a fractal object such as Koch's snowflake (Fig. 8.12) the fractal dimension $D = 1.26$.

This power law has been applied to study the tortuosity of the pathway of insect movements (Wiens *et al.* 1993). In fact, fractal dimension is a scale-independent measure of the tortuosity of a pathway

(Wiens *et al.* 1995). *D* is calculated by regressing ln (natural logarithm) of path length *C(L)* and the ln of length scale *L*, from which is subtracted 1 to yield *D*. *k* is the intercept of the regression line. When the pathway is a straight line *D* = 1 and when the pathway is so complex as to fill a plane *D* = 2. In general we can be sure that the more tortuous the pathway (high value of *D*) the more the organism interacts in fine grain with the heterogeneity of the landscape.

C(L) may be measured approximately but easily by using the grid method (see Sugihara and May 1990), which consists in superimposing a regular grid of side length *L* to the pathway or the edge of interest. At every *L* size grid the squares containing a piece of the pathway or edge are counted. Then the ln of the total number of squares is regressed with *L*. *D* is equal to the slope of the regression minus 1 (Box 8.6 and Fig. 8.14).

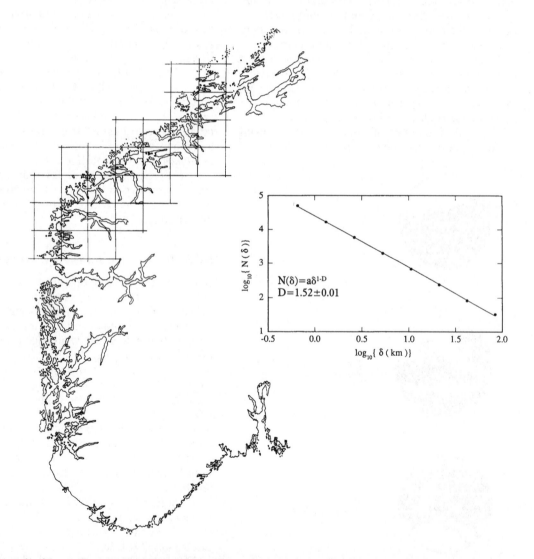

Figure 8.14 Example of the application of fractal geometry to the computation of the complexity of an edge. The coast of southern Norway, with superimposed a square grid of $\delta \approx 50$ km and the log-log plot of $N(\delta) = a\delta^{1-D}$ (from Feder 1988, with permission). Changing the value of and regressing the number of boxes $N(\delta)$ of dimension that cover the coastline with the value of δ, the fractal dimension *D* is (−1) times the slope of the regression. This method is applied in Box 8.6.

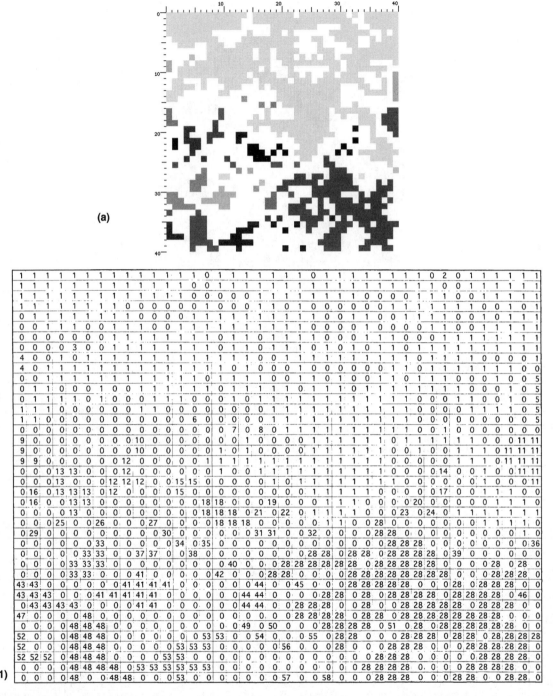

Figure 8.15 Example of application of fractal geometry to the computation of shape complexity of a land use mosaic. The maps (a) oak woodland, (b) cultivation, are from the land mosaic of Fig. 8.2 and were produced by applying the Clump routine of MacGIS. This routine renumbers each (isolated) patch with a unique number. Then, using the Export function, the numeric value is exported (a1, b1) and processed using the program listed in Box 8.6. The fractal dimension (1.61 for oak woodland and 0.95 for cultivation) is the slope of the regression log of area on log of perimeter.

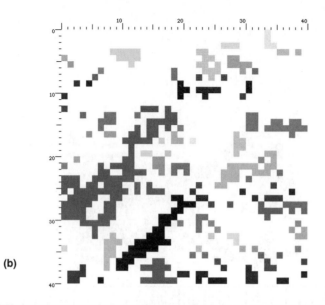

(b)

(b1)

```
 0  0  0  0  0  0  0  0  0  0  0  0  0  0  0  0  0  0  0  0  0  0  0  0  0  0  0  0  0  0  0  0  0  0  0  1  0  0  0  0  0  0
 0  0  0  0  0  0  0  0  0  0  0  0  0  0  0  0  0  0  0  0  0  0  0  0  0  0  0  0  0  0  0  0  0  0  0  1  0  0  0  0  0  0
 0  0  0  0  0  0  0  0  0  0  0  0  0  0  0  0  0  0  0  0  0  0  0  0  0  0  2  2  2  0  0  0  0  1  1  0  0  0  0  0  0
 0  0  0  0  0  0  0  0  3  3  3  3  3  0  0  0  0  0  0  0  0  0  4  4  0  0  0  0  0  0  0  0  0  0  5  5  0  6  0
 7  0  0  0  0  0  0  0  3  3  3  3  3  0  0  0  0  0  0  0  0  0  4  4  0  0  8  8  0  0  0  0  0  0  0  0  0  0  0  0
 0  9  0  0  0  0  0  0  0  3  0  0  0  0  0  0  0  0  0  0  0  4  4  0  4  0  8  8  8  8  0  0  0  0  0  0  0  0  0  0
 0  0  0  0  0  0 10  0  0  0  0  0  0  0  0  0  0  0  0 11  0  0  0  0  4  4  4  0  0  8  0 12  0  0  0  0  0  0  0  0
 0  0  0  0  0 13  0  0  0  0  0  0  0  0  0  0  0  0  0 11  0  0  0 14  0  4  0 15  0  0  0  0  0  0  0  0  0  0  0  0
 0  0 16  0 17  0  0  0  0  0  0  0  0  0  0  0  0  0 18  0  0  0  0  0  0  0  0  0  0 19  0  0  0  0  0  0  0  0  0  0
 0  0  0  0  0  0  0  0  0  0  0  0  0  0  0  0  0  0 18 18  0 20 20 20 20  0  0  0 19  0  0  0  0  0  0  0  0  0  0
21  0  0  0  0  0  0  0  0  0  0  0  0  0  0  0  0  0 18  0  0 20  0  0  0  0 22  0  0  0  0  0  0  0  0  0  0  0  0
 0  0  0  0  0  0  0  0  0  0  0  0  0  0  0  0  0  0  0  0  0  0  0  0  0  0  0  0  0  0  0  0  0  0  0  0  0  0  0
23  0  0  0  0  0  0  0  0  0  0  0  0 24 24 24  0 25  0  0  0  0  0  0  0  0  0  0  0  0  0  0  0  0  0  0  0  0
 0  0  0 26  0  0  0 27 27  0  0  0 28  0  0  0  0 25 25  0  0  0  0  0  0  0  0  0 29 29  0  0  0  0  0  0  0  0
 0  0  0  0  0  0 27 27 27  0  0  0  0 25 25 25 25  0  0  0  0  0  0  0  0  0 29 29  0 30 30  0 30 30  0
 0  0  0  0  0 27  0  0  0  0 25  0 25 25  0 25  0 31  0  0  0  0  0  0  0  0 29 29  0 30 30 30 30 30 30 30
 0  0  0 32  0  0 33 33  0  0 25 25  0 25 25  0 34  0  0  0  0  0  0 35  0  0  0  0  0  0  0 30  0  0
 0  0  0  0  0  0  0  0  0  0 25 25 25 25 25  0 36  0 34 34  0 37  0  0  0  0 35 35  0  0  0  0  0  0  0
 0  0  0  0  0 39  0  0  0 38  0 25  0  0  0  0  0  0  0  0  0 35 35  0  0  0  0  0  0
 0  0  0  0 39  0  0  0  0 25  0 40 40  0  0 41  0  0  0  0  0  0  0 35  0  0  0 42 42  0  0  0  0
 0 43  0 39 39  0  0  0 25 25  0  0 44 44  0  0  0  0  0  0  0  0  0 45  0 42  0
46  0  0  0  0 25  0 25 25 25 25  0  0 44 44  0 47 47 47  0  0  0  0  0 42 42 42 42 42  0 42 42  0  0  0  0
 0  0  0  0  0 25 25  0 25 25 25 25  0 44  0 47  0 47  0  0  0  0 42 42 42  0 42 42 42  0 48  0  0  0  0
25 25 25  0  0 25 25 25 25 25 25 25  0  0  0  0  0  0  0  0  0  0 42 42 42  0 49  0  0  0  0
25 25 25  0 25 25  0 25 25 25  0 50  0 52  0 42 42  0  0  0  0  0 53  0  0  0
25  0 25 25 25 25 25 25 25 25  0 50  0 51 51 51  0 54  0  0  0 55  0  0  0 56  0  0
 0 25 25 25 25 25  0  0 25 25  0  0  0  0  0  0  0 57 57  0  0 58 58  0  0  0  0  0 59  0 60 60 60  0  0  0  0
25 25 25 25 25  0  0 25 25  0  0  0  0  0  0  0  0 57 57 57  0 61  0  0 62  0  0  0 63  0  0 64 64 64 64
25 25 25 25  0  0  0 25 25  0  0  0  0  0  0  0  0  0 57 57  0  0  0  0  0  0  0  0 63 63 63 63  0 64  0 64
25 25 25 25  0  0 25 25 25 25  0  0  0  0  0  0 57 57 57  0  0  0  0  0 65 65  0  0  0  0 63
 0  0  0 25 25 25 25 25  0  0  0  0  0  0  0 57 57  0 66  0  0 65  0 67  0  0  0  0  0  0 68  0
 0  0  0  0  0 25  0  0  0  0  0  0  0 57 57  0  0  0  0  0 65  0  0 69  0  0  0  0  0  0  0  0 70  0 71
 0  0  0  0  0 25  0  0 72  0  0  0  0 57 57 57  0  0  0  0  0  0 73  0 74 74  0  0  0  0  0  0  0 71 71
 0 75 75  0  0  0 72 72 72  0 57 57 57 57 57  0 76  0  0  0 77  0 74  0  0  0  0  0  0  0
78  0  0  0  0  0 72 72  0 57 57 57 57 57  0  0  0  0 79  0  0  0 80  0 81  0  0  0  0  0 82  0
 0  0  0  0  0  0 72 72  0 57 57 57 57  0  0  0  0 83  0 84  0  0  0  0  0 85  0 86  0  0  0
 0  0  0  0  0  0 72  0 57 57 57  0  0 87 87  0  0  0  0  0 88  0  0  0  0 89  0  0  0  0 90
 0  0  0 91  0  0 72  0 57 57  0 87 87 87 87 87  0  0  0  0  0  0 92  0  0 93  0 89 89  0  0  0  0
 0 91 91 91  0  0  0  0  0  0  0  0 87 87  0 94 94  0 95 95  0  0  0 93 93  0  0  0  0  0 96
 0 91 91  0  0 97  0  0  0  0  0 87 87 87 87 87 87  0  0 95  0 98 98 98  0  0 93 93  0 99 99  0  0
```

8.3.4 The fractal dimension of patches

The complexity of a polygon is expressed by the relationship $P \approx \sqrt{A^D}$ (i.e. $\log P \approx 1/2 D \log A$), where P is the perimeter and A the area. For simple polygons such as circles and rectangles $P \approx \sqrt{A}$ and $D = 1$. For irregular and complex polygons the perimeter tends to fill the plane and $P \approx A$ with $D \to 2$. This relationship can be used to calculate the complexity of coastlines of various islands, or the complexity of vegetation patches using the same scale of measurement, assuming a self-similarity between islands or vegetation patches of different sizes. In this case the scale of the ruler should be small enough to avoid that with decreasing island perimeter/area the measured shapes become Euclidean ($D \to 1$).

The fractal dimension is obtained by regressing $\log(P)$ on $\log(A)$, where $D = 2*$regression slope (see Box 8.7).

Using this approach Krummel *et al.* (1987) demonstrated that fractal dimension changes according to the size of forest patches. Moving to small woodlots, produced mainly by human disturbance in a less disturbed, large forest, the fractal dimension shows an increase. This means that at a larger scale where the natural processes are dominant the landscape is more convoluted. In contrast, at a small scale the patterns are more regular and simplified and most have been produced by human disturbance regimes. In terms of fractal analysis this means that moving across scales the invariance falls into two distinct subsets. The first is dominated by human disturbance regimes, and the second by natural processes (Fig. 8.16). This approach is very interesting and can be applied to a broad range of phenomena in which shape is an important component of the ecological process.

8.3.5 Semivariance and fractal analysis

Russell *et al.* (1992) studied sea bird dispersal and the distribution of food, applying fractal analysis. In this study to assess the relationship between prey and predators the authors utilized a method based on geostatistics and regionalized variables (RV) theory (Fig. 8.17).

This regionalized variable is too complicated to be expressed by a simple mathematical model because it has a deterministic character in the near-by samples, but is stochastic in that value at a given point because it cannot be calculated from neighbouring samples.

The semivariance $g(\gamma) = 1/2N(\gamma) \Sigma(Xj - (Xj + \gamma)^2)$ is in relation with γ (that is, the sampling interval) by the relationship $2g(\gamma) = \gamma^{(4-2D)}$, where $g(\gamma)$ is the semivariance at interval γ, the fractal dimension $D = (4 - m)/2$ ($\gamma \to 0$) where m is the slope ($4 - 2D$) of $\ln g(\gamma) - \ln \gamma$ (Burrough 1981).

Fractal analysis is particularly efficient to describe variations across a wide range of scales.

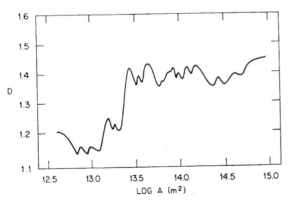

Figure 8.16 The fractal dimension changes abruptly when the change of scale of resolution reflects the change between human shaping of the landscape mosaic (small scale) to natural shaping forces (largest scale). D = fractal dimension, log A = logarithm of area investigated (from Krummel *et al.* 1987, with permission).

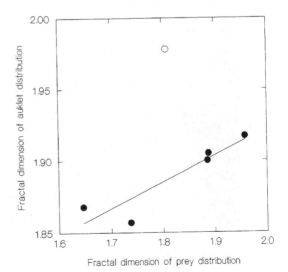

Figure 8.17 Significant relationship between the fractal dimension of predators (auklet) and prey across six transects (from Russell *et al.* 1992, with permission).

Generally the patterns are produced by a variety of processes operating at many spatial scales and levels of organization.

Fractals help us to understand how ecological processes operate, although *per se* the information cannot be correlated with a specific process. Fractal analysis has also been used for computing the home range of animals (Loehle 1990), using the box-counting method with boxes of different size to assess the pattern of occupation of space by an object. In the case presented by Loehle a radio-collared hawk visited the home range in an irregular way: some areas were more frequented but others were never visited. If we encircle the whole area as the maximum distance in which an animal has been observed we lose much detail. In contrast, considering movements at different spatial scales we can measure the complexity of the area use, from roosting places to the entire landscape.

Recently Farina (1997b) employed fractal analysis to investigate the resources available for a bird community living in a submontane ecotone of the Apennines by regressing log of perimeter with the log of area of the patches created and interpolating bird abundance (Fig. 8.18). The fractal dimension D of the patches clearly decreases with the increase in bird abundance. In this case the simpler shape of low-abundance patches seems to be due to a more sparse distribution of resources that forces birds to search in finer and more convoluted ways. In con-trast, abundant resources are more concentrated and birds can track them in a more simplified way. However, for very abundant resources D has a rising trend. In this case it is reasonable to assume that the scaling properties of these resources are changing.

8.3.6 Application of fractals to animal behaviour

If we adopt the organism-centred view of the landscape it is essential to know the perception resolution or grain and the range of scale, the extension at which an organism considers the landscape as heterogeneous.

Grain and extension can change during the development of the organism (e.g. fish size classes) and according to the seasons (e.g. migratory birds), and this can create problems when employing a too simple or too sophisticated model.

The movements of animals are easily detected and measurable for many species, and are strongly affected by the body mass of the organism and by the resolution at which the organism perceives its surroundings. Considering that a landscape is a hierarchical array of patchiness, it is important to distinguish at what resolution the organism interacts with the environment and in what way it ignores the patterns that are outside the specific range of resolution.

If we use the movements of the organism as an indicator of landscape interaction we can assume that species which move slowly perceive the environment at a finer scale than the species moving faster. But when two species are different sizes it seems impossible to compare their behaviour and use of resources because they are scaled differently. In fact, movement pathway is strongly influenced by landscape structure and by the size of the organism. The body size–scale-dependent movements are difficult to compare.

Applying fractal analysis it is possible to create a scale-independent measure of the movement because the fractal dimension of a movement pathway is scale independent and may be used to compare different taxa (Wiens *et al.* 1995). A clear example of such an approach is presented by With (1994), studying the movement patterns of three acridid grasshoppers (*Orthoptera*) in a grassland mosaic. Manipulation of species in a controlled microlandscape has been carried out by the author, confirming the predictive potentiality of such an approach to a broad-scale experimentally

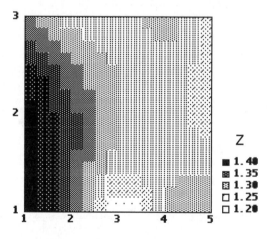

Figure 8.18 Categories of bird abundance (X), seasons (Y) (1 = fall, 2 winter, 3 breeding) and fractal dimension (Z) of log area/log perimeter per bird in a submontane ecotone (from Farina 1997b, with permission).

intractable landscape. The larger species, *Xanthippus corallipes*, moved at a faster rate than the two other species, *Psoloessa delicatula* and *Opeia obscura*, and perceived the microlandscape in a different way, presenting different values of fractal dimension D. This species had more linear movements than the two others, which had less perception of the heterogeneity of the landscape. These two species have a similar D value and this probably means that they use resources in similar ways (Fig. 8.19).

To calculate the fractal dimension D the divider method was used; this consists in measuring the total length of the pathway (summation of distances between points) at different ruler lengths. In this case 25 ruler lengths were selected. The minimum ruler length was calculated as the average distance between points and the maximum as one-third of the total path length, considering that at least three points are required for a linear regression. For instance, Wiens *et al.* (1993) found in three tenebrionid beetles living in prairie in northeastern Colorado a similar D dimension of pathway convolution, differing significantly from 1 (linear movement) and from random walk; these three species demonstrate a similar strategy across scales.

Fractal analysis has been used to study stress in *Capra pyrenaica* caused by pregnancy and by *Sarcoptes scabies* infection (Alados *et al.* 1996) (Fig. 8.20).

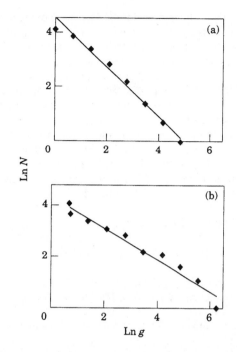

Figure 8.20 Log-log plot of the cumulative frequency (*N*) of feeding gaps greater than or equal to an interval *g* in *Capra pyrenaica* (a) in parasitized, (b) in non-parasitized females (from Alados *et al.* 1996, with permission).

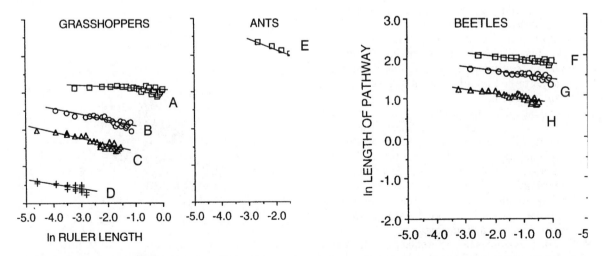

Figure 8.19 Regression showing the relationship between the pathway length and the measurement scale in grasshoppers, ants and beetles. Grasshoppers: A = *Opeia obscura* nymph.; B = *Opeia obscura* adult; C = *Psoloessa delicatula*; D = *Xanthippus corallipes*; Ants: E = *Pogonomyrmex occidentalis*; Beetles: F = *Eleodes extricata*; G = *Eleodes obsoleta*; H = *Eleodes hispilabris* (from Wiens *et al.* 1995, with permission).

Owing to the increase in metabolic rate due to the infection stress, the infected animals show a reduction in the complexity of exploratory behaviour. It is well known that the complexity of biological tissue decreases under the effects of pathologies. In the same manner behaviour suffers a reduction in variability. Head-lift behaviour, which consists in the interruption of feeding behaviour and in lifting the head as antipredatory vigilance, was analysed by regression after a log-log linearization of the frequency of head-lifts, where $F(\Delta t)$ is the frequency of head-lifts at intervals of duration Δt $F(\Delta t) = k(\Delta t)^D = k(1/\Delta t)^{-D}$, where k is constant and D is the fractal dimension.

In Figure 8.20 the log-log regression of $F(\Delta t) \times \Delta t$ shows a reduction of the fractal dimension between non-parasitized females (a) and parasitized females (b).

The feeding gap frequency N with frequency G greater than or equal to a determined size g fits a power law $N(g \geqslant G) = kg^{-B}$, where k is constant, $g = 1, 2, ...2^n s$ and B is the Korcak–Mandelbrot exponent.

8.3.7 Summary

- Patch size, shape and spatial arrangement are important characteristics in landscape analysis. These attributes affect processes, patterns and organisms in different ways.
- To measure these landscape attributes several indices are available, but it is difficult always to find a correlation between the adopted indices and the ecological information.
- A short review of these indices has been carried out at two levels: at patch scale and at landscape scale. In the first case some simple indices have been described to collect attributes such as area, perimeter and patch shape.
- At landscape scale descriptors of mosaic complexity indices, such as richness, number of patches, relative importance, Shannon diversity of patches (type and size classes), dominance, evenness, entropy and contagion seem very important.
- The use of distance and texture measurements offers new possibilities to investigate the complexity of landscape mosaics.
- Practical examples are offered and small routines in Basic are provided to calculate the main patch and landscape attributes.
- Lacunarity as a measure of landscape structure is presented as an introduction to fractal geometry.

This last argument has been extensively presented as a powerful tool to measure the complexity of natural patterns. The vastness of the fractal approach has obliged us to discuss only a narrow set of applications to the field of landscape ecology.

- Fractal dimensions of edges, the shape of patches and animal movement patterns have been described through simple and effective indices.

8.4 GEOGRAPHIC INFORMATION SYSTEMS

8.4.1 Introduction

Geographic information systems (GIS) are a technology for handling spatial data. Developed in recent years, they are now applied in many fields from local to global scale. The GIS is a configuration of computer hardware and software for the capture, storage and processing of spatial information, both numerical and qualitative, creating and updating maps, combining and interpreting maps; they are a revolution in map structuring, content and use. GIS can be classified according to application, addressed as urban information systems, spatial decision support systems, soil information systems, planning information systems, land information systems etc.

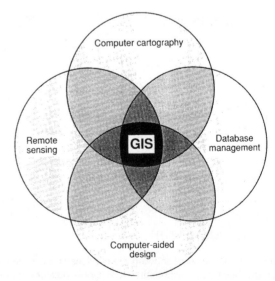

Figure 8.21 GIS is the combination of different procedures and methodologies (from Maguire 1991, with permission).

Box 8.6

PROGRAM TO CALCULATE THE FRACTAL DIMENSION OF EDGES USING THE DIVIDER METHOD
This routine has been prepared in Basic

```
REM MS1=Number of columns
REM MS2=Number of rows
REM Patch=Value of grid cell : 0 or 1
REM NODIV=Number of divisions
REM Divider=box dimension
REM CODE=Number of boxes for each divider

PRINT "Matrix size: (x)";:INPUT MS1
PRINT "Matrix size: (Y)";:INPUT MS2

DIM patch(MS1,MS2), per(500), tboxl(500), dividerl(500), tboxll(500)

INPUT "file name"; pop$
Y=1
OPEN "I", #1, "hard disk:"+pop$

100 IF EOF(1) THEN GOTO 1000
X=X+1

INPUT #1,a
patch(Y,X)=patch(Y,X)+a

IF X=MS1 THEN Y=Y+1:X=0

GOTO 100

1000'

PRINT "The matrix is :":PRINT MS1, MS2

PRINT "Select the Number of dividers":INPUT NODIV
FOR M=1 TO NODIV

PRINT "Select the size of divider":INPUT divider
divider(M)=divider

CODE(M)=MS1/divider(M)
NEXT M

FOR j=1 TO NODIV
```

```
PRINT j, divider(j), CODE(j)
NEXT j

FOR j=1 To NODIV
DIM tot(CODE(j),CODE(j))
tt=0
FOR Y=1 TO MS1
FOR X=1 TO MS2
If patch(Y,X)><0 THEN cody=INT((Y−1+divider(j))/divider(j)):
      codx=INT((X−1+divider(j)):'PRINT y,cody,x,codx
IF patch(Y,X)><0 THEN tot(cody,codx)=tot(cody,codx)+1

NEXT X
NEXT Y

FOR l=1 TO CODE(j)
FOR k=1 TO CODE(j)
CC=CC+1

IF tot(l,k)><0 THEN tt=tt+1
NEXT k
NEXT l
PRINT "cc";CC, tt
ERASE tot

tbox(j)=tbox(j)+tt

tt=0
CC=0

NEXT j

FOR j=1 TO NODIV
PRINT j, divider(j), tbox(j)
tboxl(j)=tboxl(j)+LOG(tbox(j))
dividerl(j)=dividerl(j)+LOG(divider(j))

'Computation of regression

sumx=sumx+tboxl(j)
sumy=sumy+dividerl(j)
sumxy=sumxy+tboxl(j)*dividerl(j)
sumxsq=sumxsq+tboxl(j)*tboxl(j)
sumysq=sumysq+dividerl(j)*dividerl(j)
NEXT j
xbar=sumx/NODIV
Ybar=sumy/NODIV

FOR l=1 TO NODIV
```

```
diffX=diffX+(xbar–tboxl(l))*(xbar–tboxl(l))
diffY=diffY+(Ybar–divider(l))*(Ybar–dividerl(l))
NEXT I

sqxy=sumxy–(sumx*sumy)/NODIV

b=1–(sqxy/diffY)

PRINT "FRACTAL DIMENSION"; b

END
```

Box 8.7

PROGRAM TO CALCULATE THE FRACTAL DIMENSION *D* OF A LAND MOSAIC

This routine has been prepared in Basic

```
REM PROGRAM TO CALCULATE THE FRACTAL DIMENSION
REM REGRESSING LOG PERIMETER ON LOG AREA
REM data are from matrix produced by the routine "Clump" of MacGIS

INPUT "File name:"; pop$
INPUT "Matrix size: columns"; Ncol
INPUT "Matrix size: rows"; Nrow
INPUT "Cell length:"; length

DIM patch(Nrow,Ncol), per(Nrow*Ncol), area(Nrow*Ncol), perl(Nrow*Ncol),
areal(Norw*Ncol)
y=1
OPEN "I", #1, pop$

100 IF EOF(1) THEN GOTO 1000
x=x+1

INPUT #1, a
patch(y,x)=patch(y,x)+a

IF x=40 THEN y=y+1:x=0

GOTO 100

1000'
FOR y=2 TO Nrow
```

```
FOR x=2 TO Ncol
IF patch (y,x)><0 THEN GOSUB 3000
NEXT x
NEXT y

FOR y=1 TO Nrow
FOR x=1 TO Ncol
area(patch(x,y))=area(patch(x,y))+(length*length)
NEXT x
NEXT y

FOR l=1 TO Nrow*Ncol
IF per(l)><0 THEN cc=cc+1
IF per(l)><0 THEN perl(cc)=perl(cc)+LOG(per(l)): areal(cc)=areal(cc)+LOG(area(l))

NEXT l
PRINT cc

FOR l=1 TO cc
PRINT #2, area(l); ",";per(l)

sumx=sumx+areal(l)
sumy=sumy+perl(l)
sumxy=sumxy+areal(l)*perl(l)
sumxsq=sumxsq+areal(l)*areal(l)
sumysq=sumysq+perl(l)*perl(l)
NEXT l

xbar=sumx/cc
ybar=sumy/cc

FOR l=1 TO cc
diffx=diffx+(xbar-areal(l))*(xbar-areal(l))
diffy=diffy+(ybar-areal(l))*(ybar-areal(l))
NEXT l
sqxy=sumxy-(sumx*sumy)/cc

b=sqxy/diffY

PRINT "FRACTAL DIMENSION"; b

END

3000'
REM length of patch perimeter
IF patch(x+1,y)=0 THEN per(patch(x,y))=per(patch(x,y))+length
IF patch(x-1,y)=0 THEN per(patch(x,y))=per(patch(x,y))+length
IF patch (x,y+1)=0 THEN per(patch(x,y))=per(patch(x,y))+length
IF patch(x,y-1)=0 THEN per(patch(x,y))=per(patch(x,y))+length
RETURN
```

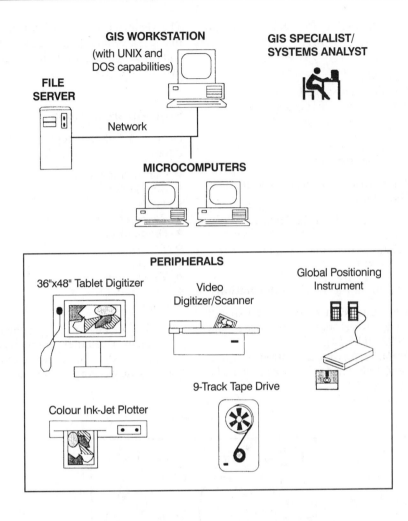

Figure 8.22 Components of a GIS laboratory. The complexity and the costs of the hardware and of the driving software are extremely variable. Often it is necessary to have a large central computer memory to harvest digital images. Different software is necessary to couple with GIS software, especially for manipulating remote sensing data (from Coulson *et al.* 1991, with permission).

The spatial information represented by localization in a geographic space of the attributes of an event can today be easily handled and processed thanks to the combination of spatial statistics, mathematical procedures and computer software. The combination of these three components creates a geographic information system (Burrough 1986) in which computer cartography, database management, remote sensing procedures and computer-aided design are the structuring components (Maguire *et al.* 1991) (Fig. 8.21). GIS are used by a growing number of people in different fields, from economics to social science and planning. The

dimensions, the software and the technical requirements (hardware) range from a few kilobytes to many gigabytes, with costs from a few dollars to several thousand. Some are extremely user-friendly, but others need a dedicated operator (Fig. 8.22).

The incredible development of these systems makes it difficult to describe the applications available, which are rooted in geography and computing.

In landscape ecology GIS are a fundamental tool, especially if used as a platform to manipulate models and real data, transferring information from implicit to explicit analysis.

GIS appear indispensable for most landscape investigations, e.g.:

- Land use change;
- Vegetation patterning;
- Animal distribution across the landscape;
- Linking remote sensing with topography;
- Modelling processes across the landscape.

8.4.2 The information process

Three types of information on landscape features are considered by a GIS:

- Names and characteristics of the features;
- Their locations;
- The spatial relationship to one another.

8.4.3 Representation of the spatial information

Two systems are available to represent the spatial information as lake–forest–field maps: raster and vectorial (Figs 8.23 and 8.24).

The raster format is the representation of a feature using discrete units known as pixels or cells.

The vectorial format utilizes the position of point, line and area and their connectivity. The advantages and disadvantages of both systems are summarized in Table 8.8.

8.4.4 GIS organization

Data in a GIS are represented by a hierarchy of information that has at the highest level:

- Cartographic models
- Map layers
- Titles
- Resolutions
- Orientations
- Zones
- Labels
- Values
- Locations
- Coordinates.

A GIS operates in geographical or virtual space in which every element (in both raster and vectorial format) is indicated by two coordinates (X,Y) and consequently every computational procedure is

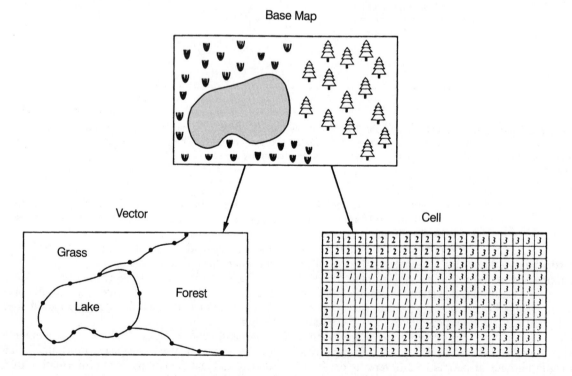

Figure 8.23 Representation of raster and vectorial formats of a map (from Coulson *et al.* 1991, with permission).

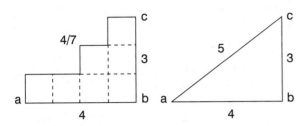

Figure 8.24 Raster versus vectorial representation of reality. The real measures in raster format are coded and some biases in estimation have to be considered (from Burrough 1986, with permission).

available either inside a layer or map or between different layers (Fig. 8.25).

8.4.5 Cartographic model

The cartographic model is a collection of map layers and describes a well defined area. The map layers contain information on the same geographic area, such as size, name of localities, history of special sites etc. Each cartographic model has an implicit and an explicit form.

8.4.6 Map layer

Often indicated simply as a layer, this is a conventional map reporting a variety of information, not necessarily geographical. In a map layer information is represented by occupied cells described according to different attributes of the subject. A layer contains other explanatory information, e.g. title, resolution, orientation and zone(s).

- Title, e.g. vegetation, bird richness, is important when layers are manipulated and each layer entered into an algebraic formula.
- Resolution of a layer represents the relationship between the ground and 'on paper' distance.
- Orientation of a layer indicates the relationship between geographic and cartographic direction.

Table 8.8 Comparison of raster and vectorial format. (From Burrough 1986, with permission)

Vector methods
Advantages
    Good representation of phenomenological data structure
    Compact data structure
    Topology can be completely described with network linkages
    Accurate graphics
    Retrieval, updating and generalization of graphics and attributes are possible
Disadvantages
    Complex data structures
    Combination of several vector polygon maps or polygon and raster maps through overlay creates difficulties
    Simulation is difficult because each unit has a different topological form
    Display and plotting can be expensive, particularly for high quality, colour and cross-hatching
    The technology is expensive, particularly for the more sophisticated software and hardware
    Spatial analysis and filtering within polygons are impossible

Raster methods
Advantages
    Simple data structure
    Overlay and combination of mapped data with remotely sensed data are easy
    Various kinds of spatial analysis are easy
    Simulation is very easy because each spatial unit has the same size and shape
    The technology is cheap and is being energetically developed
Disadvantages
    Volumes of graphic data
    The use of large cells to reduce data volumes means that phenomenologically recognizable structures can be
        lost and there can be a serious loss of information
    Crude raster maps are considerably less beautiful than maps drawn with fine lines
    Network linkages are difficult to establish
    Projection transfomation is time consuming unless special algorithms or hardware are used

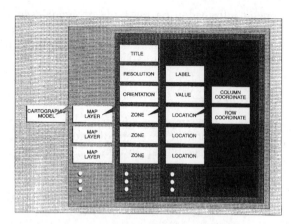

Figure 8.25 Scheme for hierarchical organization of geographic data (from Tomlin 1990, with permission).

- Zones are a part of the map distinguished by certain attributes from other zones, for example a forest patch, a field, urban area etc. Each zone is indicated by a label, i.e. a name, and values, which can be a further specification of the zone: for example, a zone labelled 'forest' with a value of 200 means 'forest 200 m from road'. Value can be expressed as a ratio, interval, ordinal or nominal. Nominal is the representation of a quality of the zone, e.g. as 'dense shrubland'.
- Location is the elementary unit of a map. In raster format this is represented by a square cell or grid cell, and by pixels in image processing.
- Coordinates are a pair of numbers expressed in geographic units, for example the distance in metres from the Equator and Greenwich, or simply by an 'on paper' numerical scale.

8.4.7 Procedures for cartographic handling and modelling

Most of the GIS available today have in common the capacity to manipulate information both at a single location and at the scale of the entire map. All data in a GIS occupy a precise position according to true geographical or working coordinates (x,y). In the first case this may be UTM (kilometric coordinates).

8.4.8 Capturing data

Data can be captured in many ways, the least expensive being to rasterize existing maps such as topo-

graphic maps, land cover maps etc. Data may already be in raster format if they are provided by remote sensing satellite stations.

Digitization is a very precise but expensive procedure. For example, the location with coordinates x_{ij} y_{ij} in which i is the attribute (e.g. land cover type) and j is the layer, can be manipulated algebraically adding the same location in other layers x_{ij} y_{ij} + x_{im} y_{im} (where m is another layer). Other procedures can manipulate entire layers, e.g. *thismap + thatmap = new map*.

8.4.9 Some cartographic modelling procedures

According to Tomlin (1990) at least three operations are available in GIS procedures:

- Local operations: For each location it is possible to associate a new value that represents the transformation of the same value applying mathematical functions to each location's value(s) on one or more existing map layers.
- Zonal operations: For each location a new value is computed as a function of the existing value from a specified layer.
- Focal operations: For each location a new value is computed according to the character of the neighbouring locations of the same map layer, or on other map layers (Fig. 8.26).

8.4.10 Commands in GIS

Many routines are available to transform data. For local operations arithmetic menus are available (see, e.g., MacGis program; Hulse and Larsen 1989) which have functions such as:

add: adds values of two or more existing maps;
average: averages values of two or more existing maps;
cover: covers values of one existing map with one or more existing maps;
divide: divides values of one existing map by one or more existing maps;
maximize: maximizes values of two or more existing maps on new maps;
minimize: minimizes values of two or more existing maps on new map;
multiply: multiplies values of two or more existing maps;
subtract: subtracts values of one existing map from one or more existing maps.

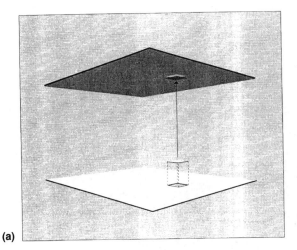

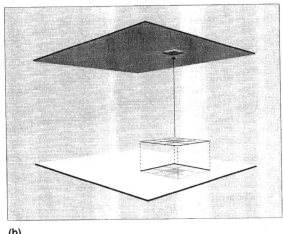

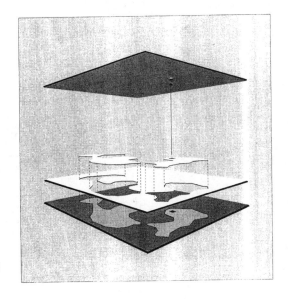

Figure 8.26 The fundamental capabilities of a GIS can be categorized into three groups of procedures: (**a**) functions of a single value; (**b**) functions within neighbourhoods; (**c**) functions of entire zones (from Tomlin 1990, with permission).

For focal operations a neighbourhood menu is available with functions such as:

clump: generates a map of contiguous like-valued cells;

differentiate: generates slope map from surface data;

interpolate: calculates intermediate values from two positions (Fig. 8.27);

orient: generates aspect map from surface data;

radiate: generates viewshed map from specified viewer locations;

scan: classifies neighbourhood of specified locations as to neighbouring values (Fig. 8.28);

score: compares and summarizes values of two maps on a point-by-point basis;

smooth: generates map of surface from map of contours;

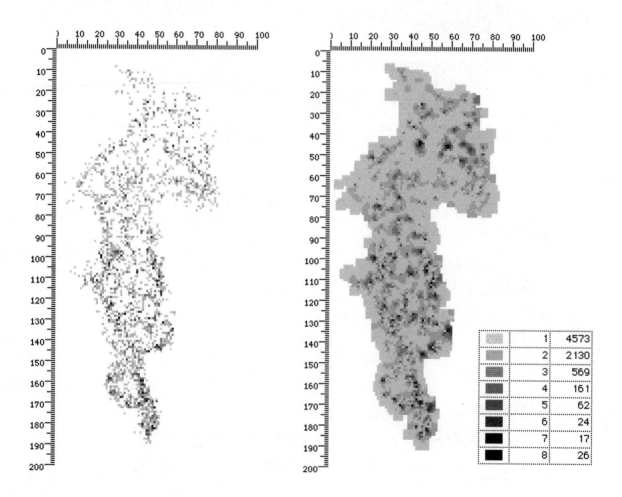

Figure 8.27 Example of application of Interpolate command to transform point counts of annual bird abundance in a surface map of abundance in the upland ecotones of Logarghena. The geographical position of birds was recorded using a GPS (from Farina 1997b, with permission).

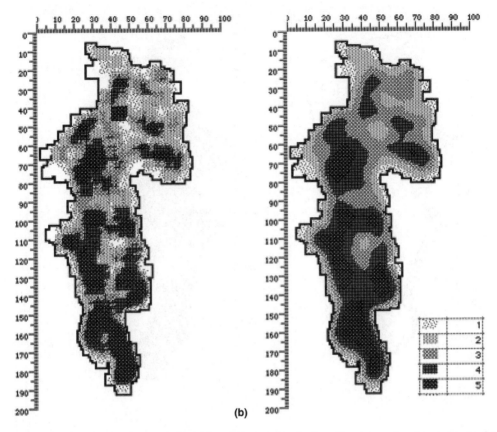

(a)   (b)

Figure 8.28 Change of resolution in a map of bird abundance during the breeding season in a Logarghena upland ecotone (Farina 1997b) from 20 m to 200 m applying the routine MacGIS scan with a radius of 10 cells around every cell and adding all abundance values according the subroutine total, and recording each cell according to the total abundance found in that radius. The new map (**b**) shows a coarse pattern of abundance that in some cases seems more useful for comparison with environmental attributes.

spread: generates map of proximity to specified locations (Fig. 8.29).

8.4.11 GIS and remote sensing

Remote sensing information produced by satellite or aerial photography must be interpreted before they can be used in a GIS. Generally, data from remote sensing are imported into a GIS after classification and georeferencing.

The procedures of land classification are independent of GIS techniques, but when the data have to be introduced into a GIS it is necessary to know at least the spatial scale (the resolution) of the images for a georeference.

8.4.12 Scaling in GIS

In landscape ecology it is often useful to process spatial data at different scales; in fact, in a landscape patterns and processes are visible and function along a broad range of spatial scales (Turner *et al.* 1989). Only a limited number of routines are available to carry out these procedures. Baker and Cai (1992)

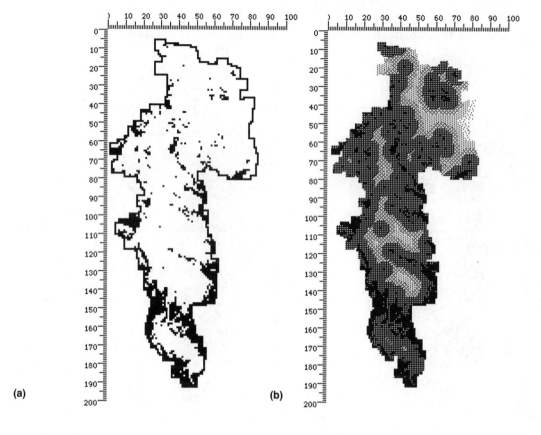

Figure 8.29 Example of application of Spread command to classify the upland ecotones of Logarghena according to the distance from tree cover. (a) Distribution of trees; (b) distance from trees (five classes) (from Farina 1997b, with permission).

presented a program operating in GRASS able to calculate more than 60 routines on landscape structure (e.g. distance, size, shape, fractal dimension, perimeters, diversity, texture, juxtaposition, edges etc.), with different possibilities of sampling areas of several sizes, changing 15 scales of analysis or using a moving window (Fig. 8.30).

See also Fig. 8.31 for cell resolution of a map.

8.4.13 Key study 1: Ecotope classification, application of cartographic modelling to the Aulella watershed (Farina 1996)

The classification of a landscape into its component ecotopes is a common procedure in landscape analysis. The ecotope represents a piece of land in which some characteristics are homogeneous and distinct from neighbouring areas. These are often

a combination of physical and biotic components. For example, we need to localize all the cultivated areas situated in hilly conditions (Apennines range, northern Italy, Mediterranean basin) between a range of 200 and 500 m a.s.l., in which the slope orientation pertains to SE, S and SW quadrants. In this case the ecotope should represent the xerophitic, warmer site in which olive orchards and vineyards are cultivated, which for convenience we will call 'wintering.site'. In Figure 8.32 a picture of such an ecotope is given as an example.

This ecotope plays a relevant role in the wintering of many small birds (Farina 1987) and its localization can assume major importance, for example when protection action has to be planned. The starting point is the input of basic information, in this case:

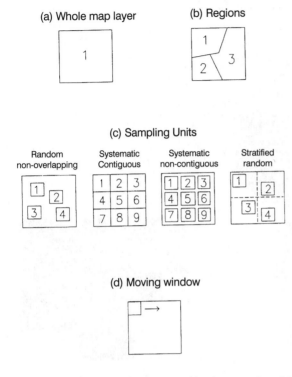

(a) Whole map layer   (b) Regions

(c) Sampling Units

Random non-overlapping   Systematic Contiguous   Systematic non-contiguous   Stratified random

(d) Moving window

Figure 8.30 Different sampling combinations employed in the r.l programs (from Baker and Cai 1992, with permission).

- Elevation map (topography);
- Aspect map (aspect.map);
- Land use map (cultivation).

Topographic and vegetation maps were created by digitizing existing maps; the aspect map was created using the 'orient' procedure of a MacGIS routine on an elevation map. The model is very simple and is obtained using the following steps:

1. Select all cells ≥ 200 and <500 from *topography* map and save a new map as *hill*.
2. Select from *land use* map cultivation, and save a new map *cultivation*.
3. Run 'orient' routine on *topography* and save a new map *orient*.
4. Select from 'orient' values 4, 5, 6, which represent SE, S and SW orientation respectively, and save a new map *southern*.
5. Recode, assigning value 1 to *southern*, *cultivation* and *hill*, and save each map assigning the name *southern.1, cultivation.1, hill.1.*

6. Use the routine 'add' to add the three maps, then select only location with value 3; this represents the ecotope 'wintering.site' that we have searched (Fig. 8.33).

This procedure can be simplified and the chronological sequence changed without modifying the results.

8.4.14 Key study 2

The aim of this example is to classify a soil surface in terms of the potential connectivity of perennial grass cover (*Brachypodium geneuense*) for the movement of micromammals such as *Pitymys savii*.

In this example a small surface area of mountain prairie (Logarghena, northern Apennines) has been captured using a digital camera (Quick Take) suspended from a balloon and switched remotely at an altitude of 30 m (Fig. 8.34a). The image was classified using Dimple software according to nine land cover types, from unclassified to annual grassland, to perennial grassland and shrubs. The image classified was imported into MacGIS using the 'import' routine; the ground resolution was 0.25 cm/pixel.

The category of perennial grassland was isolated (Fig. 8.34b) and then, using the MacGIS 'spread' routine, the distance of each cell from this target land cover was calculated (Fig. 8.34c). Ten distance categories were distinguished, from 1 to 10, using darker grey tones (Fig. 8.34d). This last map may be used as a basal map for modelling the probability of *Pitymys savii* finding a non-hostile habitat when moving to forage.

8.4.15 Summary

- GIS are a powerful tool in landscape ecology.
- Although some GIS are expensive and data input time consuming, an intelligent selection of software allows us to gain most of the benefits of a more complete and sophisticated system.
- GIS procedures allow us to store, manipulate and transfer field and remote sensing information rooted in geographical gridding.
- Handling spatial data guarantees precision in the information, but also a powerful system of handling data. In fact, cartographic modelling allows us to overlap, combine and separate selected information from the different layers that compose a GIS, as illustrated in the two key examples.

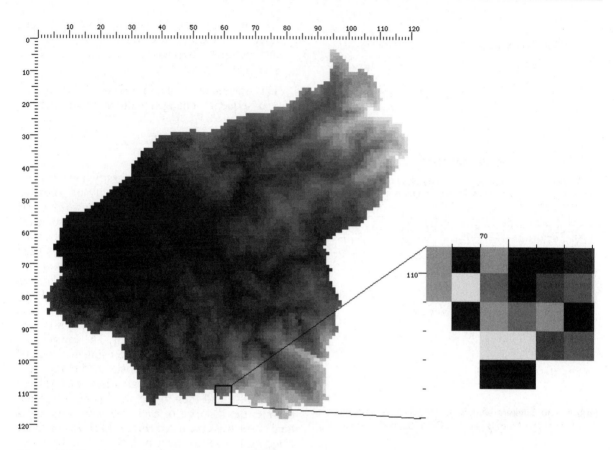

Figure 8.31 Resolution is the dimension of the grid cell or pixel composing the matrix. In this case a 24 km area (Aulella watershed) has a resolution of 200×200 m (adapted from Farina 1997a).

- We consider it extremely useful to couple different routines, such as remote sensing and image interpretation, with GIS procedures.

8.5 REMOTE SENSING IN LANDSCAPE ECOLOGY

8.5.1 Introduction

The goal of this section is to review briefly the application of remote sensing techniques to the study of the landscape. For more information on these techniques, which are becoming increasingly popular and sophisticated, see Johnson (1969).

Remote sensing and GIS are often used together in landscape analysis and classification (e.g. Davis and Goetz 1990). For example, the detection of linear structures in a human-dominated landscape can represent an important element for landscape description. Goossens *et al.* (1991) have tested the capacity of SPOT multispectral imaging of a rural area in Belgium to detect ecological infrastructures (Fig. 8.35).

Linear structures such as hedgerows, edges, drainage ditches and roads are detected in different measure by SPOT sensors. The height of objects (tall objects have longer shade), the sun's angle (the lower the angle the longer the shade), and the orientation of trees and edges play a role in edge detection.

The availability of images captured by satellites or aircraft along a multispectral reflectance permits a wide range of applications of these techniques to landscape study.

Remote sensing data can be collected at different spatial and temporal scales, creating a powerful tool

Figure 8.32 View of olive orchards growing on warm, hilly slopes of the northern Apennines (Lunigiana). Their functional characters are homogeneous enough to be considered a distinct ecotope in this landscape mosaic.

to study processes. Remotely sensed data can be processed and improved through digital techniques, available on inexpensive platforms (e.g. Dimple, Multispec). The resolution scale ranges from 10 to 20 or 30 m. The Landsat Thematic Mapper has a resolution of 30×30 m on the ground; SPOT has a 10×10 m ground resolution.

In a landscape each element has a characteristic multispectral response.

Hall *et al.* (1991) used Landsat Multispectral Scanner (MSS) data (60 m resolution) to study the boreal forest landscape across a temporal scale of 10 years.

8.5.2 Effects of sensor spatial resolution on landscape structure parameters

Remote sensing has been usefully applied to verify the scaling properties of the landscape (Nellis and Briggs 1989, Benson and MacKenzie 1995). For example, Nellis and Briggs (1989) used band rationing between the Landsat Multispectral Scanner (MSS) and Thematic Mapper (TM) combined with image textural features at three scales. Band rationing was used to estimate and monitor green biomass (Fig. 8.36). Healthy vegetation reflected 40–50% of incident near-infrared energy (0.7–1.1 μm) and absorbed 80–90% of the incident energy in the visible (0.4–0.7 μm) part of the spectrum close to the red region (0.6–0.7 μm). The ratios between MSS bands 4–2 and TM band 4–3 (the ratio of near-infrared energy to visible energy) were particular useful in determining the quantity of biomass or net primary production.

Three scales of resolution were used: 5 m (digitally improved aerial photographs) (spectral resolution 0.3 μm), 30 m (TM) and 80 m (MSS). The textural contrast between adjacent pixels was carried out using 0–255 categories of contrast.

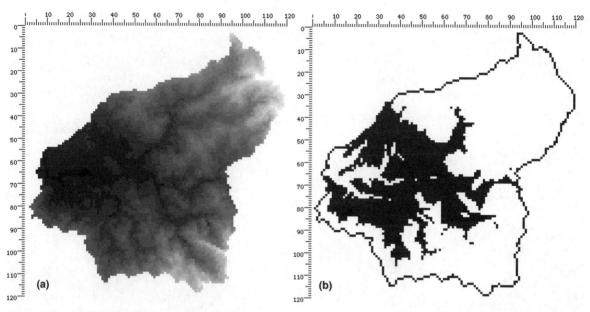

Figure 8.33a Data preparation 1: Selection from the topographic map (**a**) of the hilly portion of terrain ranging from 200 to 500 m of Aulella watershed (Hill.1) (**b**) (data from Farina 1997a).

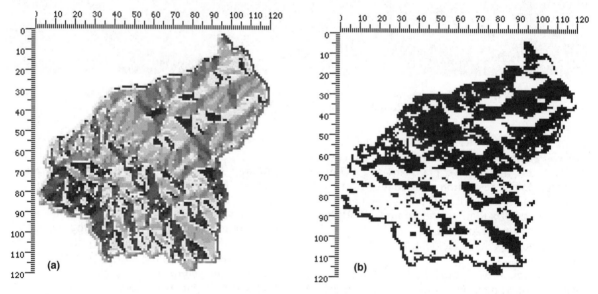

Figure 8.33b Data preparation 2: Selection of SE, S, SW and W aspects of Aulella watershed slopes from orient map (**a**) in southern map (southern.1) (**b**) (data from Farina 1997a).

Different components of the Konza prairie landscape reacted differently to sensor resolution. Areas of dense patchiness have to be analysed at a finer scale than the areas burned every four years. In this case aerial photography and Landsat were preferred to study the unburned areas. Using high-resolution

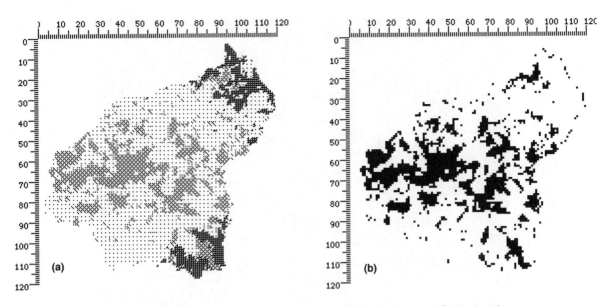

Figure 8.33c Data preparation 3: Selection (**b**) from the land cover map (**a**) of cultivations (Cultivation.1).

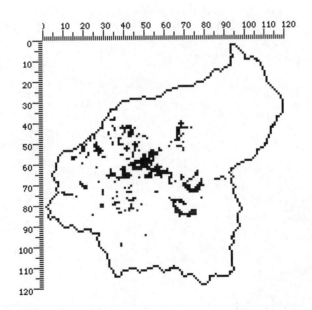

Figure 8.33d Calculation of ecotope wintering site adding maps hill.1 + southern.1 + cultivation.1 after their recording with 1 for all non-0 values. The cells with value 3 have a combination of all three maps, therefore are the selected ecotope (data from Farina 1997a).

digitally enhanced photography it was possible to appreciate great differences in the degree of textural contrast. In conclusion, the Landsat MSS digital data are less suitable for mapping the Konza prairies.

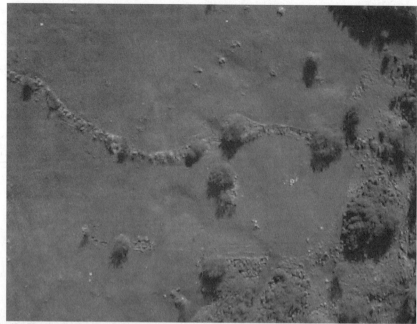

(a)

(b)

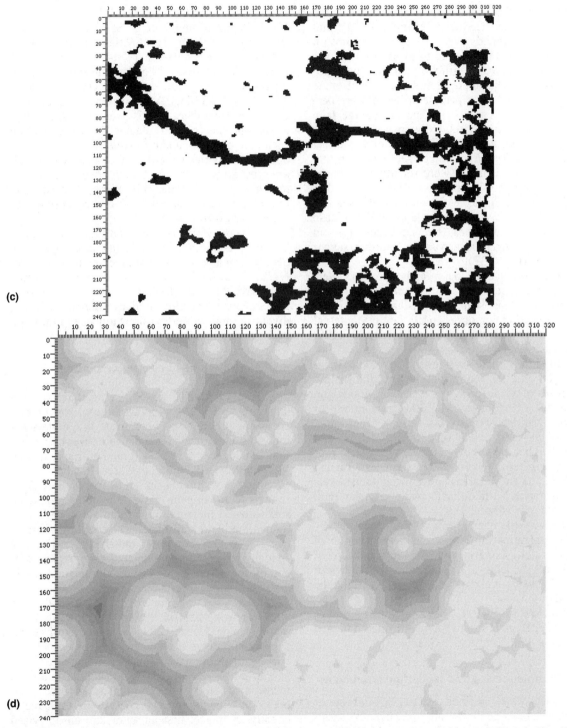

Figure 8.34 (**a**) Low-altitude image of Logarghena prairies taken by a digital camera suspended from a balloon (spring 1996); (**b**) the same image after classification into nine classes of land cover; (**c**) separation of *Brachypodium genuense* cover; (**d**) map of distance (10 categories) from the *Brachypodium* cover (from Farina, unpublished).

SPOT Multispectral High-Resolution Visible (HRV), Landsat Thematic Mapper (TM) and NOAA, Advanced Very High Resolution Radiometer (AVHRR) were used to compare the change of landscape structure in the northern Wisconsin lake district by Benson and MacKenzie (1995). The landscape parameters used were percentage of water, number of lakes, average lake area and perimeter, fractal dimension and three measures of mosaic texture (homogeneity, contrast and entropy). For these indices see earlier in this chapter. These measures were sensitive to sensor resolution, moving from 20 m for HRV to 1100 m for AVHRR. For instance, the number of lakes and the percentage of water in the landscape decrease as grain size increases, but homogeneity and entropy were invariant across the satellite sensor resolutions (Fig. 8.37).

Considering that the major landscape parameters are sensitive to sensor resolution and that at different resolutions landscape patterns appear or disappear, it would be important to fill the great interval resolution between the TM and AVHRR (30–1100) using interpolating procedures.

Scaling the remote sensing land cover classification is an efficient way of determining the scaling properties of a landscape. For instance, Moody and Woodcock (1995) used MT 30 m resolution for Plumas National Forest (California), aggregated into five classes of resolution (90, 150, 240, 510 and 1020 m). The changes in land cover type due to progressive aggregation procedures were tested using five independent variables: mean patch, mean interpatch distance, Shannon index, variance mean ratio and initial proportion.

Generally the decrease in resolution caused by pixel aggregation introduces a large proportion of errors, which strongly influence the reliability of large-scale modelling. Understanding the behaviour of aggregated spatial data is useful in preserving the information.

Simmons *et al.* (1992) used satellite images to evaluate ecological scale applying the methods used by Carlile *et al.* (1989) (see Chapter 3).

8.5.3 Remote sensing and landscape boundaries

Boundaries, also defined as transitional zones between landscape units, are fundamental structures in any landscape (see Chapter 5). Metzger and Muller (1996) propose a new boundaries index using remote sensing data (Table 8.9).

Using a Landsat Thematic Mapper image of Jacaré-Pepira basin (State of Sao Paulo, southeast Brazil) a supervised classification was made using red (TM3), near infrared (TM4) and shortwave-infrared (TM5) spectral bands with the normalized difference vegetation index (NDVI = (TM4–TM3)/(TM4 + TM3)). After the land cover classification the boundary pixels were extracted (Fig. 8.38a) and then dilated, placing the same pixel on each side of the boundary pixels (Fig. 8.38b). The dilated boundaries were then added and the boundary type extracted for each land cover as a combination of the three types (Fig. 8.38c). For example, the value 6 is the result of 2 + 4 land cover types; 12 is the result of 8 + 4, etc.

This approach clearly showed the importance of boundary analysis in interpreting landscape complexity. Landcover boundary diversity is significantly related to land cover shape. Elongated riparian units have the highest value for boundary diversity and covert proportion indices. Shape, richness, diversity and covert proportion are important descriptors of landscape complexity.

8.5.4 Forest ecology and remote sensing

The first civilian earth-observing satellite was launched in 1972. From that time to the present enormous progress has been made in remote sensing, especially in the field of land cover classification. Owing to the coarse grain of Landsat TM, MSS and AVHRR, only large areas can be usefully classified. For this reason landscape ecology finds in remote sensing an invaluable tool, producing sophisticated information in natural as well as in human-modified landscapes. The coupling of image analysis to GIS facilities, and then incorporating spatially referenced data such as topography, have greatly improved the quality of the numerical and graphical output (Fig. 8.39 and Table 8.10).

Mostly used in land cover classification, the remote sensing approach is moving towards landscape change. Because this technique is of recent origin comparison is restricted to two decades, but future perspectives are strongly predicted.

The procedure to classify a satellite multiband image is described in detail in many software handbooks, but two different approaches are available to classify an image: first an automatic unsupervised classification based on differences in the spectral characters of pixels. After the classes have been generated it is necessary to assign meaning to them

Table 8.9 Indices used to calculate boundary proportion and complexity. (From Metzger and Muller 1996, with permission)

A Indices of land cover and boundary proportion

$pi = Ai/A$ Proportion of land cover i where Ai is the area of land cover i and A the area of landscape

$qi = Bi/B$ Proportion of land cover boundary i where Bi is the boundary area of land cover i, B is the landscape boundary area

$Fi = Bi/Ai$ Shape index or proportion of boundary area in land cover i

$F = B/A$ Landscape fragmentation index or proportion of boundary area within the landscape

B Indices of land cover boundary complexity

$Ci = Bci/Ai$ Proportion of convergency points (or coverts) in land cover i where Bci is the area of coverts in land cover i

NBi Number of boundary types in land cover i

$$HB = \sum_{k=1}^{Nbi} -q_{ki} * \log_2 q_{ki}$$ Boundary diversity index in land cover i of each boundary type k

C indices of landscape boundary complexity

$C = Bc/A$ Proportion of convergency points (or coverts) in landscape where Bc is the area of coverts in the landscape

NB Landscape boundary richness index i.e. sum of the number of simple contacts (points where two land covers converge) and coverts (points where three or more land covers converge)

$$HB = \sum_{k=1}^{NB} q_k * \log_2 q_k$$ Landscape boundary diversity index where q_k is the boundary area proportion in the landscape of each boundary type k

(converting classes to land cover type). For example, the sensor most accurate in forest mapping seems to be the TM sensor, especially bands 1, 5 and 7.

The second method, called supervised classification, consists in the creation of training sets, selecting aggregations of pixels on which can be recognized a distinct land cover or vegetation type. The image will be classified according to the selected training sets.

8.5.5 Landscape classification using remote sensing

Remote sensing allows us to classify land cover types fairly reliably. However, higher-order patterns

in the land cover mosaic, representing different landscape types, cannot be immediately recognized using this technique. Haines-Young (1992) has coupled TM and MSS digital processes of selected areas in southeast Wales to the TWISPAN program. He grouped the land cover combination 1×1 km cells of the National Grid into landscape classes, finding a good correspondence with ITE (Institute of Terrestrial Ecology) Land Classes for Great Britain and the Agricultural (Jane) Census statistics for England and Wales (Fig. 8.40). To classify the landscape the classified images were geometrically

Table 8.10 Main satellite types, spatial resolution and temporal coverage

System		IFOV	Repeat coverage
SPOT	Multispectral	20 m	Days-variable
	Panchromatic	10 m	Days-variable
LANDSAT	MSS	80 m	Several days
	TM	30 m	Several days
NOAA	AVHRR	~1 km	Few hours
METEOSAT		~2.5 km	30 minutes

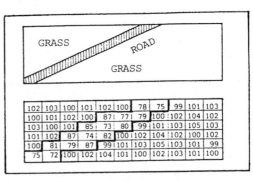

Figure 8.35 Schematic representation of a road as detected by SPOT sensors. The numbers refer to digital numbers (DNS) (from Goossens *et al.* 1991, with permission).

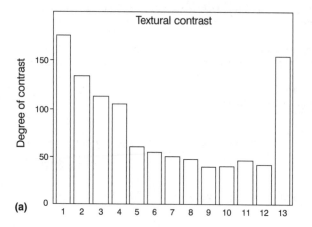

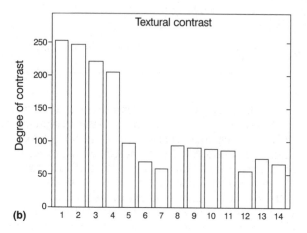

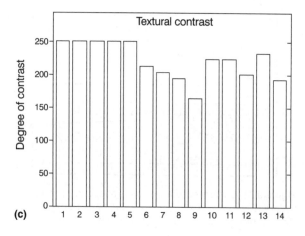

corrected using a nearest-neighbour algorithm to the UK Ordnance Survey National Grid. In this way it is possible to include landscape type in the general framework of remote sensing.

8.5.6 Calibration centre concept

The combined use of TM and AVHRR was tested by Iverson *et al.* (1994) at two locations (Illinois and Smoky Mountains). The TM data were used to classify a smaller area (calibration centre) into forest/non-forest. The combination of TM and AVHRR data allows a better interpretation of AVHRR data (Fig. 8.41).

Unsupervised classification techniques were used to classify forest/non-forest TM data at a resolution of 30 m and the classification was verified using aerial photographs. The AVHRR pixels (1 km resolution) were superimposed on to TM pixels (30 m resolution): 154 AVHRR pixels, each of which contained 1369 TM pixels for the Illinois area, and 99 AVHRR pixels with 871 TM pixels for the Smoky Mountains. Regression analysis was carried out to determine the best correlation between spectral characteristics of the AVHRR data and TM data.

AVHRR on its own can underestimate or overestimate forest cover; the TM calibration improves

Figure 8.36 The degree of textural contrast using (**a**) digitally enhanced aerial photographs, (**b**) Landsat TM, (**c**) Landsat MSS. (**a**) Digitally enhanced aerial photography. 1. Forest, north N1B; 2. Forest, south N1B; 3. Forest, east UB; 4. Forest, west UB; 5. N1B, upland unburned grassland; 6. N1B, lowland unburned grassland; 7. NUB, lowland unburned grassland; 8. 10B, lowland unburned grassland; 9. 4G, lowland unburned grassland; 10. 10B, lowland unburned grassland; 11. UB, burned grassland; 12. Grazed grassland, south Konza; 13. Interstate highway. (**b**) Landsat thematic mapper. 1. Forest, King Creek; 2. Forest, Shane Creek; 3. Forest, brome, Kins Creek; 4. Forest, brome, Shane Creek; 5. N1B, unburned grassland; 6. 4B, unburned grassland; 7. CU, unburned grassland; 8. 2DE, burned grassland; 9. N484, burned grassland; 10. N1A, unburned grassland; 11. ID, unburned grassland; 12. 3B3UD, burned grassland; 13. Fallow, agriculture. (**c**) Landsat multispectral scanner. 1. Forest, King Creek; 2. Forest, Shane Creek; 3. Forest, brome, Kins Creek; 4. Forest, brome, Shane Creek; 5. 4G, unburned grassland; 6. N1A, unburned grassland; 7. CU, unburned grassland; 8. 2DE, unburned grassland; 9. N484, burned grassland; 10. 4B, burned grassland; 11. ID, burned grassland; 12. 3B3UD, burned grassland; 13. Cropland, agriculture; 14. Fallow, agriculture (from Nellis and Briggs 1989, with permission).

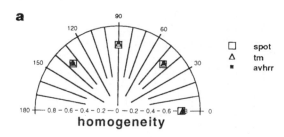

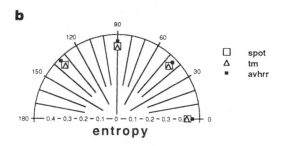

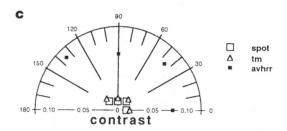

Figure 8.37 Texture values for HRV, TM and AVHRR according to angle of analysis for three texture parameters, (**a**) homogeneity, (**b**) entropy and (**c**) contrast. Only contrast shows sensitivity for sensor resolution. The angle used for proximity of pixels seems to have no effect (from Benson and MacKenzie 1995, with permission).

the discriminatory efficiency of AVHRR. The efficiency of this methodology largely depends on the soil topography and on the reflectance of vegetation. Conifer and broadleaf forests have different behaviours and calibration must be repeated for the two conditions to avoid misclassification.

8.5.7 Summary

- Remote sensing is used extensively in landscape ecology analysis.

- The classification of landscape patterns such as heterogeneity and ecotones appears especially useful.
- Landscape classification using remote sensing seems a very promising approach.
- The combination of data from different sensors can allow a better understanding of the hierarchical structure of the land mosaics and the scaled characters.

8.6 GLOBAL POSITIONING SYSTEMS (GPS)

8.6.1 Introduction

The global positioning system is a satellite-based positioning system operated by the US Department of Defense (DoD). GPS allows the collection of information about the geographical position of any location using a network of satellites. When fully operative the system consists of 24 NAVSTAR (Navigation Satellite Time And Ranging) satellites in 12-hour orbits at an altitude of 20 200 km. The system was started in 1973 and the first satellite launched in 1978. It became available for civilian users –at a high price – in 1983. The system is now less expensive, depending on the accuracy requested. The development of this technology, especially on receivers, allows centimetre precision working, even beneath dense tree coverage.

The GPS system works by using satellite trilateration, measuring the distance between satellites, calculating accurate timing, knowing where a satellite is in space and correcting ionospheric and tropospheric delays. Its accuracy depends on the equipment and processing methods used. For security reasons the DoD can degrade the accuracy with selective availability (SA). SA degrades the information used by civilian navigation receivers, miscalculating their position by hundreds of metres. To correct this bias *a posteriori* differential correction can produce an accuracy from 2 to 5 m (Fig. 8.42).

Another source of inaccuracy is produced by antispoofing (A/S), which affects dual-frequency receivers. The differential correction for SA is a technique that employs two receivers: one basic station and one (or more) remote receivers or rovers. The position of the basic station is precisely known, so it is possible to correct the biases introduced by SA. Commercial software is available to carry out this procedure.

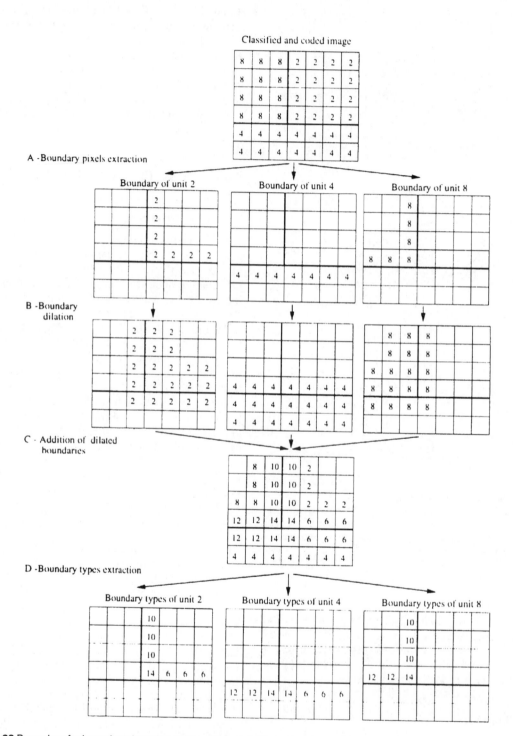

Figure 8.38 Procedure for boundary detection and topology definition using three land covers (2, 4, 8) (from Metzger and Muller 1996, with permission).

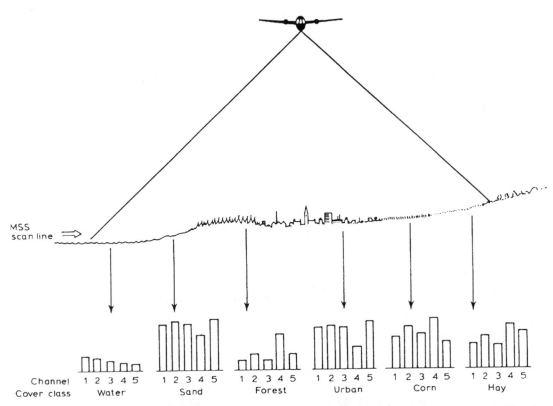

Figure 8.39 Spectral attributes belonging to different land cover types. Band 1, blue; band 2, green; band 3, red; band 4, near-infrared, band 5, thermal-infrared (from Lillesand and Kiefer 1987, with permission).

To operate a rover needs a large portion of sky visible and the position of satellites is calculated with constellation geometry. The constellation is a group of three or four satellites that are used by the receiver to calculate a position. Four satellites are necessary for a 3D position (latitude, longitude, altitude and time); for a 2D position (latitude, longitude and time) three satellites are necessary.

The quality of a position is indicated by the DOP (dilution of precision) value that appears on the receiver. The DOP value depends on the geometry of the satellites, and there are four types of DOP:

- Position (PDOP): horizontal and vertical measurements (lat, long, alt);
- Horizontal (HDOP): horizontal measurements (lat, long);
- Vertical (VDOP): altitude;
- Time (TDOP): clock offset.

A PDOP below 4 gives excellent positioning, between 5 and 9 acceptable, more than 9 poor.

It is possible to set the receiver to store only positions below a PDOP value, which is generally fixed at 6. Some receivers may require a PDOP mask with a value below 4 to achieve submetre accuracy.

To achieve the best results it is important to know the satellite position in advance. An almanac can record automatically from satellites on the receiver (basic station or rover). In this way some software can calculate for any position the value of PDOP and/or HDOP, VDOP and TDOP (Figs 8.43 and 8.44).

8.6.2 The use of GPS in landscape ecology

GPS has great potential in landscape ecology, as in many other geography-related disciplines. GPS may be used directly in the field, in cars and aero-

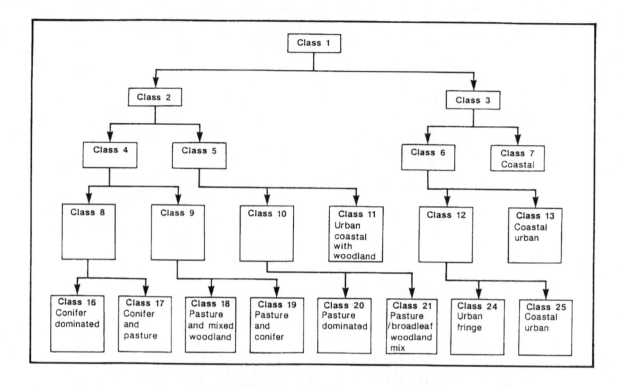

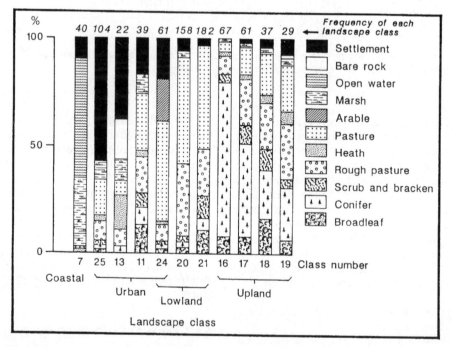

Figure 8.40 TWINSPAN classification of West Glamorgan study area, according to four levels. The first level discriminates between rough grazing, scrub and broadleaf woodland on the one hand (class 2) and areas of settlement and/or marsh (class 3) on the other (from Haines-Young 1992, with permission).

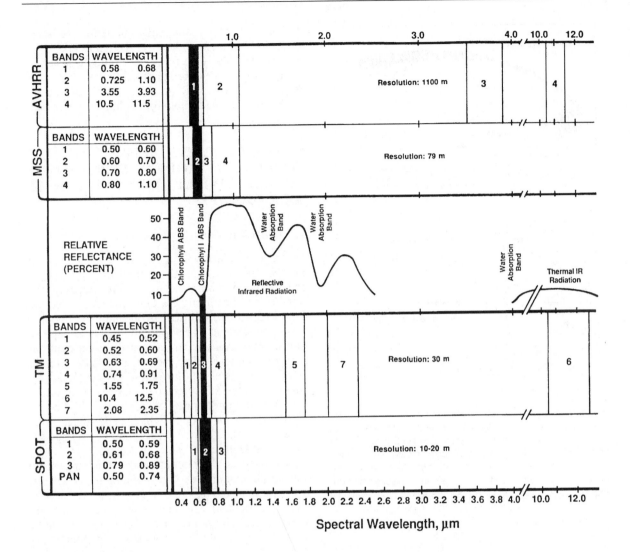

Figure 8.41 Spectral and spatial resolutions for the most common civilian satellites, AVHRR, MSS, TM and SPOT, and the electromagnetic spectral response curve for green vegetation (from Iverson *et al.* 1989, with permission).

planes and in helicopters for collecting point, line and surface features. A data dictionary prepared *ad hoc* for each project can be used in the data logger to facilitate data input. Extensively employed in forest and agricultural mensuration, it has recently been used for capturing the position of animals (Farina 1994, 1997a,b).

The geographical position of an event, such as a bird encounter, a bird call or other behaviour, is recorded at the same time as any features we care to add. This information is then transferred to a desktop computer and, using specific software, the differential correction and other processes are carried out.

Data can be exported in different formats and for any purpose. The map in Figure 8.45 was created recording birds at approximately a biweekly lag along a random transect in a montane ecotone (Logarghena prairies, northern Apennines).

After processing data can be handled in a GIS, each location can be measured as distance from other features, and different files can be merged together with an automatic rescaling of the video images.

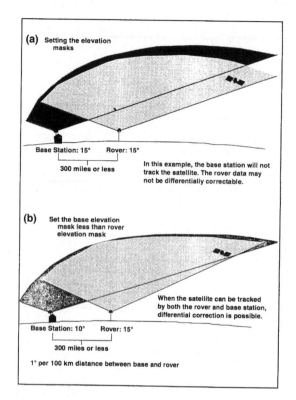

(a) Setting the elevation masks

Base Station: 15° Rover: 15°

300 miles or less

In this example, the base station will not track the satellite. The rover data may not be differentially correctable.

(b) Set the base elevation mask less than rover elevation mask

When the satellite can be tracked by both the rover and base station, differential correction is possible.

Base Station: 10° Rover: 15°

300 miles or less

1° per 100 km distance between base and rover

Figure 8.42 Representation of the interconnections between base station and rover to allow a differential correction. In (**a**) the elevation mask of the base station and of the rover is the same and the base station cannot track the satellite. In (**b**) the elevation mask of the rover is narrower than the mask of the base station. In this case the satellite can be tracked by both the rover and base station. Elevation mask is the elevational angle above the horizon (from Trimble Navigation 1994a, with permission).

The GPS methodology has been a revolution in the field of bird community ecology, boosting the census techniques, and is independent of the low resolution of topographic maps. New receivers can actually receive satellite signals under a dense vegetation cover, thereby expanding the range of possible applications.

GPS can be used to rectify aerial photographs, low-altitude oblique photographs, and for ground mapping of vegetation patches with an accuracy from a few centimetres to 5 m after differential correction.

8.6.4 Summary

- The global positioning system is a satellite-based positioning system operated by the US Department of Defense.

- This technology is usefully applied in landscape ecology to locate objects in the landscape.
- It is particularly useful to trace field maps at high resolution and to track animal movements.
- Coupled with GIS it is a powerful tool to describe the geography of ecological processes.

8.7 SPATIALLY EXPLICIT POPULATION MODELS (SEPM) TO DESCRIBE POPULATION PATTERNS IN A LANDSCAPE

8.7.1 Introduction

When a population in a landscape is reduced in number and confined to a small area the risk of local extinction may be high. Unfortunately, this condition is becoming common in fragmented and human-disturbed landscapes. To study a model of spatial movement of a population in a landscape we can apply a cellular automaton that is a cartesian grid of identical cells, each of which has a finite number of states or attributes. In the model the updating process is referred to all cells according to the state of neighbouring cells. In landscape ecology the cartesian grid, in which cells represent an area of the land surface, cell states correspond to landscape features.

To study animal populations in a heterogeneous landscape spatially explicit population models are very promising. These models incorporate the complexity of the real-world landscape (topological and chorological components). One use for these models is to investigate the responses of organisms to a broad scale of ecological processes. Changes in land use and climate modify the environmental conditions with which organisms are faced, and these processes create complex situations that cannot be studied with the traditional techniques.

A spatially explicit approach is very efficient for modelling large-scale disturbances such as fire in the ecosystem. To apply these models efficiently a strict relationship is necessary between modellers and field ecologists (Fig. 8.46).

A spatially explicit model is structured in such a way that the precise location of each element (organism, population, habitat patch) is known in comparison to the landscape features (corridors, edges, woodlots, rivers, field, forests etc.). Every element in these models can be compared with the changes that have occurred in landscape pathway composition and spatial arrangement.

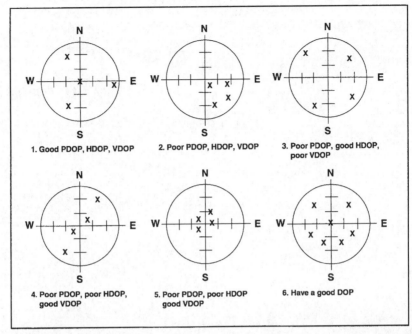

Figure 8.43 Geometry of satellite constellation and dilution of position (DOP) values for position (PDOP) (longitude, latitude and altitude), horizontal (HDOP) (latitude and longitude), vertical (VDOP) (altitude) and time (TDOP) (clock offset). The most accurate satellite geometry is illustrated in position 1, when a satellite is directly overhead and the other three are evenly spaced around the horizon (from Trimble Navigation 1994a, with permission).

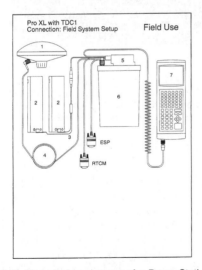

Figure 8.44 Technical equipment of a Rover Station ProXL with TDC1. 1 = Compact dome antenna; 2 = camcorder batteries; 3 = dual battery cable; 4 = antenna cable; 5 = TDC1/ProXL multiport cable; 6 = 12-channel ProXL receiver; 7 = TDC1 4MB data collector; ESP = external sensor port cable; RTCM = Radio Technical Commission for Marine Services, realtime differential correction cable (from Trimble Navigation 1994b, with permission).

To build such models it is necessary to define the grain size of the landscape (individual patch cell); habitat patches may have the dimension of a cell but generally are composed of more cells with the same characters. The extent of landscape considered spans generally from the micro- to the mesoscale (e.g. <1 ha to 104 km).

The type of landscape used in these models can be real or artificial. In the first case few land uses or other characters are preferable. The artificial landscape is used to simulate the response of species.

Models can be individual based or population based. In the first case the position of each individual annually or daily is simulated. At annual scale breeding, dispersal and mortality are considered. At daily scale foraging, growth, predator avoidance and roost selection can be monitored.

The population models are generally applied to invertebrates and mammals that have consistent populations. In this case metapopulation models can be incorporated in SEPM. SEPM have the capacity to incorporate the dispersal rules of individuals across a landscape, although the species-specific perception of the landscape is little known.

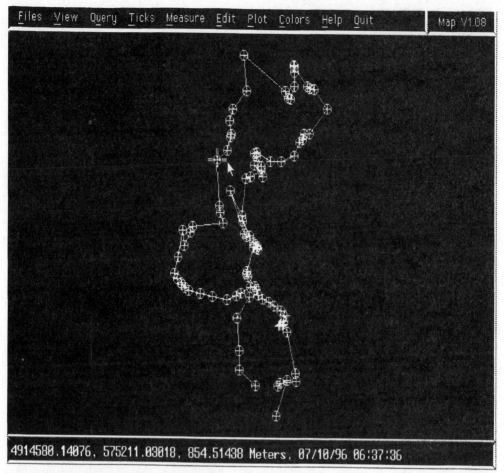

Figure 8.45 Example of output of field data (bird positions) and transect lines collected by a GPS rover along a random transect in mountain prairies (Logarghena, 44 23′ N, 9 56′ E). Data displayed are from transect no. 50 of 10 July 1996. At the bottom of the figure is the menu command of Pfinder software©. Every position is indicated by a crossed circle. The line indicates the transect. Arrow indicates the point for which, by using the menu Query, northing (UTM), easting (UTM), elevation in metres, date and local time are displayed at the bottom of figure throughout (Trimble Navigation 1994) (from Farina 1997b, with permission).

Other difficulties can emerge when the scale is enlarged and the sensitivity of a species to some character of the landscape can change.

In a real landscape the distribution of resources changes over time and space, and organisms react to this. The impossibility of manipulating a large landscape can be overcome by using organisms living in micro-landscapes that can be easily manipulated.

SEPM can be used to design natural reserves and to predict the persistence of a species according to the landscape element. These models, parameterized according to a species' life history, can verify the adequacy of existing reserves.

SEPM can serve as a bridge between spatial ecology and population genetics, exploring gene movements into a population (metapopulation). Another interesting perspective is the combination of models to predict forest dynamics due to global change, together with the reaction of animals to these environmental changes. SEPM can also include non-biological parameters such as forest economics.

For instance, the model proposed by Darwen and Green (1996) considers no landscape obstacle to population diffusion at the start of simulation (see Fig. 8.46). Although the model simulated the available space around the population, the absence of

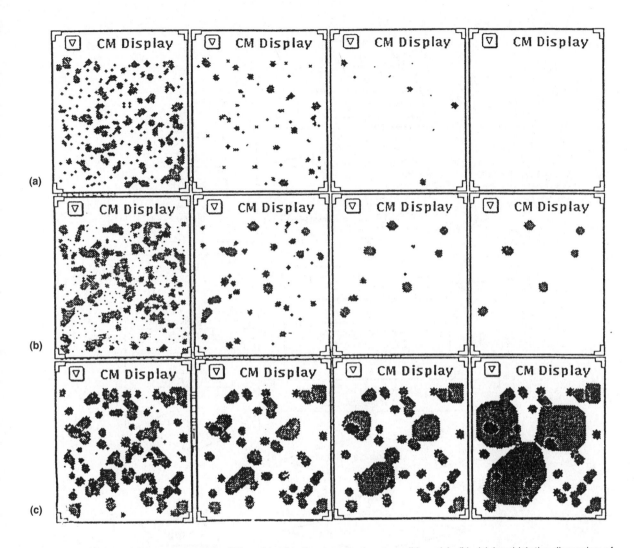

Figure 8.46 The model based on a 128×128 matrix of cells presents three conditions (**a**), (**b**), (**c**) in which the dimension of patches is increasing. From left to right illustrates the different states during running of the model. In (**a**) the patches are too small to survive; in (**b**) the patches are big enough to survive but have no chance of expanding. In (**c**) the patches are large enough to persist and to expand indefinitely (from Darwen and Green 1996, with permission).

predators and the lack of any other cause of extinction, the extinction occurred from the model. This represents a warning relevant to species conservation: if the population is too small and occupies a small area, it suffers a high risk of extinction.

8.7.2 A spatial patch dynamic model

Wu and Levin (1994) have proposed a model based on spatial patch dynamics to study dynamics at local and landscape scale (Fig. 8.47); this is useful for studying the age- and size-structured disturbance patch population and to assess how local disturbance and patch dynamics affect vegetation patterns at landscape scale. These authors studied the Jasper Ridge serpentine grassland. This area, because of the particular conditions of soil poor in Ca:Mg ratio and with a high concentration of heavy metals, has a specific grass and forb cover with a very low abundance of non-native plants. The pocket gopher (*Thomomys bottae* Mewa) is the major disturbance source. This species burrows in the soil

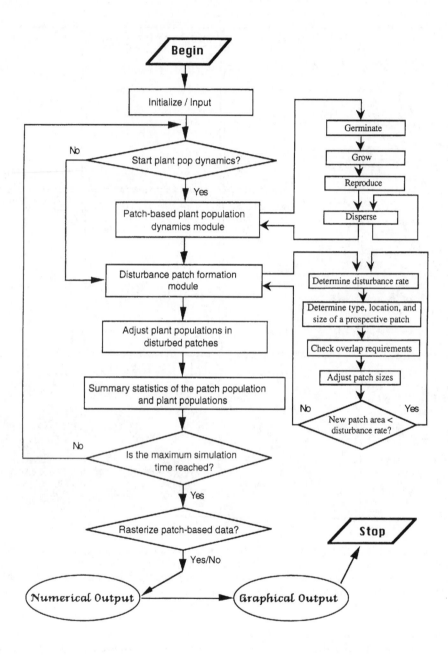

Figure 8.47 Flow chart of the spatial dynamic model PatchMod (from Wu and Levin 1994, with permission).

and creates mounds of bare soil. Most of the gopher mounds are created in April and July and about 20% of the entire area is turned over by gopher activity each year. This mound can be compared with forest gaps in developing a patch dynamic.

The recovery phases are: nudation, dispersal and colonization, plant establishment, intraspecific competition, achievement of the predisturbed state. The grassland landscape is represented by a patch mosaic of gopher mounds of different age. This

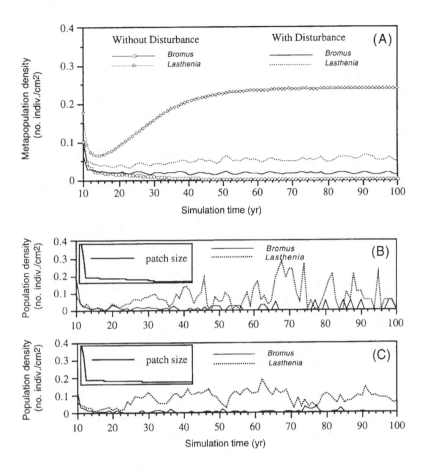

Figure 8.48 Simulation dynamic from 10 to 100 years (**a**) at landscape scale and (**b,c**) at patch scales of two competing species, *Bromus mollis* and *Lasthenia californica*. **b** and **c** are chosen as arbitary examples. In case **a** low or absent disturbance and competitive exclusion is evident, but in the presence of disturbance (gopher mounds) the local population shows high fluctuations for both species. New disturbance patches favour the persistence of *Bromus mollis* because this species has a good dispersal (large portion of seeds compared to *Lasthenia californica*). This phenomenon can also be observed in Mediterranean upland grasslands mown and grazed. The disturbance coupled with enhanced species diversity reduces the competitive exclusion effect (from Wu and Levin 1994, with permission).

model allows us to test the conceptual framework of patch dynamics, focusing on spatial heterogeneity, transient dynamics and relationships between hierarchical levels.

The model is composed of two parts: a disturbance patch demographic model and a multispecies population dynamic model. The first component considered shape, size and types of patch, their spatial distribution and the disturbance rate. The multiple-species population dynamic model took into account a patch-based multiple-species plant population dynamic model and the effect of patch age in plant demography parameters.

Pulliam *et al.* (1992) have adopted MAP (Mobile Animal Population), a spatially explicit population model to study the habitat preferences of Bachman's sparrow (*Aimophila aestivalis*), a threatened bird living in pine woods in the southeastern United States. This species is particularly sensitive to habitat management.

Three categories of variables have been incorporated in this model:

- Landscape variables that take into account habitat abundance and spatial arrangement of habitat patches;

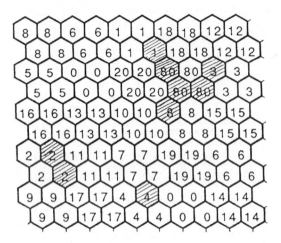

Figure 8.49 Hexagonal cells comprise a simulated landscape in which the numbers in the cells indicate the age of the stand (1–20 year harvest rotation). 80 = mature stand. Shaded hexagons are occupied by Bachman's sparrow. Hexagonal cells are used because densely packed bird territories have an approximately hexagonal shape and an extensive border that allows us to model dispersal in all directions (from Pulliam *et al.* 1992, with permission).

- Habitat-specific demographic variables (reproductive success and survival rate);
- A behavioural variable that describes the dispersal habit of the species.

The model was built using 20 age classes, old-growth category and clearcut. Simulation was made over a period of 105 years (five harvest rotations). The patterns of abundance and distribution of a population may reflect not only the current landscape characteristics, but also those of the past landscape.

Other examples of spatially explicit models have been presented by Liu *et al.* (1995). These authors have elaborated a model called ECOLECON. This is an ecological–economic model simulating animal population dynamics and economic revenue according to different forest landscape structures and timber management. It is a second-generation model built on BACHMAP, a spatially explicit model of population dynamics of Bachman's sparrow.

8.7.3 Summary

- Spatially explicit population models are importat tools to describe population patterns in a space (landscape).

- The spatial arrangement of edges, woodlots, rivers, fields and forest is introduced in the models.
- Applications of this methodology have been extended to patch dynamics, to economic interactions between population dynamics and landscape structure.

SUGGESTED READING

Burrough, P.A. *Principles of geographic information systems for land resources assessment.* Clarendon, Oxford, 1986.

Cracknell, A.P. and Hayes, L.W.B. *Introduction to remote sensing.* Taylor and Francis, London, 1993.

Diggle, P.J. *Statistical analysis of spatial point patterns.* Academic Press, London, 1983.

Feder, J. *Fractals.* Plenum, New York, 1988.

Hastings, H.M. and Sugihara, G. *Fractals. A user's guide for the natural sciences.* Oxford University Press, Oxford, 1993.

Hofmann-Wellenhof, B., Lichteneger, H., Collins, J. *Global positioning system, theory and practice,* 2nd edn. Springer Verlag, Vienna, New York, 1993.

Johnson, P.J. *Remote sensing in ecology.* University of Georgia Press, Athens, GA, 1969.

Leick, A. *GPS Satellite surveying.* Wiley and Sons, New York, 1990.

Lillesand, T.M. and Kiefer, R.W. *Remote sensing and image interpretation*, 2nd edn. Wiley and Sons, New York, 1987.

Ludwig, J.A. and Reynolds, J.F. *Statistical ecology.* John Wiley and Sons, New York, 1988.

Maguire, D.J., Goodchild, M.F., Rhind, D.W. (eds.) *Geographical information systems.* Longman Scientific and Technical, Harlow, 1991.

Mandelbrot, B. *The fractal geometry of nature.* Freeman, New York, 1982.

Ripley, B.D. *Spatial statistics.* John Wiley and Sons, New York, 1981.

Tomlin, C.D. *Geographic information systems and cartographic modelling.* Prentice-Hall, Englewood Cliffs, 1990.

REFERENCES

Acuna, J.A. and Yortsos, Y.C. (1995) Application of fractal geometry to the study of networks of fractures and their pressure transient. *Water Resources Research* **31**: 527–540.

Alados, C.L., Escos, J.M., Emlen, J.M. (1996) Fractal structure of sequential behaviour patterns: an indicator of stress. *Animal Behaviour* **51**: 437–443.

Allain, C. and Cloitre, M. (1991) Characterizing the lacunarity of random and deterministic fractal set. *Physical Review* A **44**: 3552–3558.

Baker, W.L. and Cai, Y (1992) The r.le programs for multiscale analysis of landscape structure using the GRASS geographical information system. *Landscape Ecology* **7**: 291–302.

Barak, P., Seybold, C.A., McSweeney, K. (1996) Self-similitude and fractal dimension of sand grain. *Soil Science Society of America Journal* **60**: 72–76.

Benson, B.J. and MacKenzie, M.D. (1995) Effects of sensor resolution on landscape structure parameters. *Landscape Ecology* **10**: 113–120.

Buechner, M. (1989) Are small-scale landscape features important factors for field studies of small mammal dispersal sinks? *Landscape Ecology* **2**: 191–199.

Burrough, P.A. (1986) *Principles of geographical information systems for land resources assessment.* Oxford University Press, Oxford, UK

Carlile, D.W., Skalski, J.R., Baker, J.E., Thomas, J.M., Cullinan, V.I. (1989) Determination of ecological scale. *Landscape Ecology* **2**: 203–213.

Chen, S.G., Ceulman, R., Impen, I. (1994) A fractal-based Populus canopy structure model for the calculation of light interception. *Forest Ecology and Management* **69**: 97–110.

Coulson, R.N., Lovelady, C.N., Flamm, R.O., Spradling, S.L., Saunders, M.C. (1991) Intelligent geographic information systems for natural resource management. In: Turner, M.G. and Gardner, R.H. (eds.), *Quantitative methods in landscape ecology.* Springer-Verlag, New York, pp. 153–172.

Cracknell, A.P. and Hayes, L.W.B. (1993) *Introduction to Remote Sensing.* Taylor and Francis, Bristol, UK.

Darwen, P.J. and Green, D.G. (1996) Viability of populations in a landscape. *Ecological Modelling* **85**: 165–171.

Davis, F.W. and Goetz, S. (1990) Modeling vegetation pattern using digital terrain data. *Landscape Ecology* **4**: 69–80.

Farina, A. (1987) Autumn–winter structure of bird communities in selected habitats of Tuscany. *Bollettino Zoologia* **54**: 243–249.

Farina, A. (1994) Birds in a mountain landscape. Fragmentation in agricultural landscapes. Proceedings of the third annual IALE(UK) conference, Myerscough College, Preston, 13–14 September 1994. Edited by J.W. Dover, pp. 153–160.

Farina, A. (1997a) Landscape structure and breeding bird distribution in a sub-Mediterranean region facing land abandonment. *Landscape Ecology* (in press).

Farina, A. (1997b) Resources allocation and bird distribution in a sub-montane ecotone. In: Novak, M.M and Dewey, T.G. (eds.), *Fractal frontiers.* World Scientific, Singapore, p. 478

Feder, J. (1988) *Fractals.* Plenum, New York.

Frontier, S.A. (1987) Fractals in marine ecology. In: Legendre, L. (ed.), *Development in numerical ecology.* Springer-Verlag, Berlin, pp. 335–78.

Goossens, R., D'Haluin, E., Larnoe, G. (1991) Satellite image interpretation (SPOT) for the survey of the ecological infrastructure in a small scaled landscape (Kempenland, Belgium). *Landscape Ecology* **5**: 175–182.

Gustafson, E.J. and Parker, G.R. (1992) Relationships between landcover proportion and indices of landscape spatial pattern. *Landscape Ecology* **7**: 101–110.

Haines-Young, R.H. (1992) The use of remotely-sensed satellite imagery for landscape classification in Wales (UK). *Landscape Ecology* **7**: 253–274.

Hall, F.G., Botkin, D.B., Strebel, D.E., Woods, K.D., Goetz, S.J. (1991) Large-scale patterns of forest succession as determined by remote sensing. *Ecology* **72**: 628–640.

Haralick, R.M., Shanmugam, K., Dinstein, I. (1973) Textural features for image classification. *IEEE Transactions on Systems, Man, and Cybernetics* **SMC-3**: 610–21.

Hastings, H.M. and Sugihara, G. (1993) *Fractals. A user's guide for the natural sciences.* Oxford University Press, Oxford.

Hulse, D.W. and Larsen, K. (1989) *MacGIS 2.0, A geographic information system for Macintosh.* University of Oregon.

Hulshoff, R.M. (1995) Landscape indices describing a Dutch landscape. *Landscape Ecology* **10**: 101–111.

Ichoku, C., Karnieli, A., Verchovsky, I. (1996) Application of fractal techniques to the comparative evaluation of two methods of extracting channel networks from digital models. *Water Resources Research* **32**: 389–399.

Isaaks, E.H. and Srivastava, R.M. (1989) *An introduction to applied geostatistics.* Oxford University Press, New York.

Iverson, L.R., Graham, R.L., Cook, E.A. (1989) Applications of satellite remote sensing to forested ecosystems. *Landscape Ecology* **3**: 131–143.

Iverson, L.R., Cook, E.A., Graham, R.L. (1994) Regional forest cover estimation via remote sensing: the calibration center concept. *Landscape Ecology* **9**: 159–174.

Johnson, A.R., Wiens, J.A., Milne, B.T., Crist, T.O. (1992) Animal movements and population dynamics in heterogeneous landscapes. *Landscape Ecology* **7**: 63–75.

Johnson, P.J. (1969) *Remote sensing in ecology.* University of Georgia Press, Athens, GA.

Krummel, J.R., Gardner, R.H., Sugihara, G., O'Neill, R.V., Coleman, P.R. (1987) Landscape patterns in a disturbed environment. *Oikos* **48**: 321–324.

Lathrop, R. and Peterson, D.L. (1992) Identifying structural self-similarity in mountainous landscapes. *Landscape Ecology* **6**: 233–238.

Leduc, A. Prairies, Y.T., Bergeron, Y. (1994) Fractal dimension estimates of a fragmented landscape: sources of variability. *Landscape Ecology* **9**: 279–286.

Li, H. and Reynolds, J.F. (1993) A new contagion index to quantify spatial patterns of landscapes. *Landscape Ecology* **8**: 155–162.

Li, B.L., Loehle, C., Malon, D. (1996) Microbial transport through heterogeneous porous media: random walk, fractal, and percolation approaches. *Ecological Modelling* **85**: 285–302.

Lillesand, T.M. and Kiefer, R.W. (1987) *Remote sensing and image interpretation,* 2nd edn. Wiley and Sons, New York.

Liu, J., Dunning, J.B., Pulliam, H.R. (1995) Potential effects of a forest management plan on Bachman's Sparrows (*Aimophila aestivalis*): linking a spatial explicit model with GIS. *Conservation Biology* **9**: 62–75.

Loehle, C. (1990) Home range: A fractal approach. *Landscape Ecology* **5**: 39–52.

Loehle, C. and Li, B.L. (1996) Statistical properties of ecological and geological fractals. *Ecological Modelling* **85**: 271–284.

Maguire, D.J. (1991) An overview and definition of GIS. In: Maguire, D.J., Goodchild, M.F., Rhind, D.W. (eds.), *Geographical information systems.* Longman, Harlow, pp. 9–20.

Maguire, D.J., Goodchild, M.F., Rhind, D.W. (eds.) (1991) *Geographical information systems.* Longman, Harlow.

Mandelbrot, B. (1982) *The fractal geometry of nature.* Freeman, New York.

Mandelbrot, B. (1986) Self-affine fractal sets. In: Pietronero, L. and Tosatti, E. (eds.), *Fractals in physics.* North-Holland, Amsterdam, pp. 3–28.

Metzger, J.P. and Muller, E. (1996) Characterizing the complexity of landscape boundaries by remote sensing. *Landscape Ecology* **11**: 65–77.

Milne, B.T. (1991) Lessons from applying fractal models to landscape patterns. In: Turner, M.G. and Gardner, R.H. (eds.), *Quantitative methods in landscape ecology.* Springer-Verlag, New York, pp. 199–235.

Moody, A. and Woodcock, C.E. (1995) The influence of scale and the spatial characteristics of landscapes on land-cover mapping using remote sensing. *Landscape Ecology* **10**: 363–379.

Musick, H.B. and Grover, H.D. (1991) Image textural measures as indices of landscape pattern. In: Turner, M.G. and Gardner, R.H. (eds.), *Quantitative methods in landscape ecology.* Springer-Verlag, New York, pp. 77–103.

Nardelli, R. (1995) *Distribuzione e abbondanza del pettirosso* (Erithacus rubecula) *attraverso un gradiente ambientale del sistema appenninico tosco-emiliano.* Unpublished thesis.

Nellis, M.D. and Briggs, J.M. (1989) The effect of spatial scale on Konza landscape classification using textural analysis. *Landscape Ecology* **2**: 93–100.

O'Neill, R.V., Krummel, J.R., Gardner, R.H. *et al.* (1988) Indices of landscape pattern. *Landscape Ecology* **1**: 153–162.

Perfect, E. and Kay, B.D. (1995) Applications of fractals in soil and tillage research: a review. *Soil and Tillage Research* **36**: 1–20.

Perfect, E., McLaughlin, N.B., Kay, B.D., Topp, G.C. (1996) An improved fractal equation for the soil water retention curve. *Water Resources Research* **32**: 281–287.

Perrier, E., Mullon, C., Rieu, M. (1995) Computer construction of fractal structures: simulation of their hydraulic and shrinkage properties. *Water Resources Research* **31**: 2927–2943.

Plotnick, R.E., Gardner, R.H., O'Neil, R.V. (1993) Lacunarity indices as measures of landscape texture. *Landscape Ecology* **8**: 201–211.

Pulliam, H.R., Dunning, J.B., Liu, J. (1992) Population dynamics in complex landscapes: a case study. *Ecological Applications* **2**: 165–177.

Rasiah, V. (1995) Fractal dimension of surface-connected macropore count-size distributions. *Soil Science* **159**: 105–108.

Rex, K.D. and Malanson, G.P. (1990) The fractal shape of riparian forest patches. *Landscape Ecology* **4**: 249–258.

Romme, V.H. (1982) Fire and landscape diversity in subalpine forests of Yellowstone National Park. *Ecological Monographs* **52**: 199–221.

Russell, R.W., Hunt, G.L. Jr., Coyle, K.O., Cooney, R.T. (1992) Foraging in a fractal environment: Spatial patterns in marine predator–prey system. *Landscape Ecology* **7**: 195–209.

Shannon, C.E. and Weaver, W. (1962) *The mathematical theory of communication.* University of Illinois Press, Urbana.

Simmons, M.A., Culliman, V.I., Thomas, J.M. (1992) Satellite imagery as a tool to evaluate ecological scale. *Landscape Ecology* **7**: 77–85.

Sugihara, G. and May, R.M. (1990) Applications of fractals in ecology. *TREE* **5**: 79–86.

Tomlin, C.D. (1990) *Geographic information systems and cartographic modelling.* Prentice-Hall, Englewood Cliffs.

Trimble Navigation (1994a) *GPS mapping systems, general reference.* Trimble Navigation Ltd, Sunnyvale, CA.

Trimble Navigation (1994b) *GPS Pathfinder Series Pro XL System. Operation manual.* Trimble Navigation Ltd, Sunnyvale, CA.

Turner, M.G., O'Neill, R.V., Gardner, R.H., Milne, B.T. (1989) Effects of changing spatial scale on the analysis of landscape pattern. *Landscape Ecology* **3**: 153–162.

Turner, S.J., O'Neill, R.V., Conley, W., Conley, M.R., Humpries, H.C. (1991) Pattern and scale: statistics for landscape ecology. In: Turner, M.G. and Gardner, R.H. (eds.), *Quantitative methods in landscape ecology.* Springer-Verlag, New York, pp. 17–49.

van Dorp, D. and Opdam, P.F.M. (1987) Effects of patch size, isolation and regional abundance of forest bird communities. *Landscape Ecology* **1**: 59–73.

van Hees, W.W.S. (1994) A fractal model of vegetation complexity in Alaska. *Landscape Ecology* **9**: 271–278.

Villard, M. and Maurer, B. (1996) Geostatistics as a tool for examining hypothesized declines in migratory songbirds. *Ecology* **77**: 59–68.

Wiens, J., Crist, T.O., Milne, B.T. (1993) On quantifying insect movements. *Environmental Entomology* **22**: 709715.

Wiens, J.A., Crist, T.O., With, K.A., Milne, B.T. (1995) Fractal patterns of insect movement in microlandscape mosaics. *Ecology* **76**: 663–666.

With, K.A. (1994) Using fractal analysis to assess how species perceive landscape structure. *Landscape Ecology* **9**: 25–36.

Wu, J. and Levin, S.A. (1994) A spatial patch dynamic modeling approach to pattern and process in an annual grassland. *Ecological Monographs* **64**: 447–464.

Glossary and acronyms

A/S Antispoofing. A P-code (precise code) from a GPS satellite for military use that cannot be received for non-military use

Algorithm A set of rules to produce a computation

Almanac Information transmitted by satellites describing the orbit of the GPS satellite

Alpha diversity The number of species in a collection

Area-sensitive species Species sensitive to habitat size, that require a large stand of the same type

Autoecology The ecology of a species

Autopoietic capacity The capacity of a system to self-organize and to maintain a creative attitude across homoeostatic and homeorhetic responses to changing conditions

AVHRR Advanced very high resolution radiometer

Bajada A broad continuous alluvial slope extending from the base of a mountain range to an inland basin in semiarid and desert regions, as in SW USA

Base station The GPS reference station in which geographical coordinates are known with precision (see differential correction)

Basic Beginner's All-purpose Symbolic Instruction Code. A simple computer programming language, generally used by inexperienced users

Beta diversity The rate of change in a species along a gradient from one habitat to another

Bighorn A wild sheep, *Ovis canadensis*, living in the western North America

Biodiversity The number of species present in a site, the variety of living organisms

Bit map A sequence of bits (i.e. 0/1) on the grid

Buffer Transitional area acting as a filter or a mitigator of disturbance processes

Cadastral map Maps generally at scale of 1:2.000 that describe the bounds of properties, roads and hydrographic nets

Cantor dust An infinite number of points scattered over an interval after an infinite number of operations. This is produced starting from a unit of which a generator removes at each step the open middle third.

Climax communities More or less stable communities at the terminal stage of ecological succession

Cluster A group of cells or pixels connected to each other

Coarse grained When a pattern or a mosaic has large components

Connectedness The physical distance from elements of the same type. Used generally for forest patches

Connectivity Functional attribute of connectedness

Constellation The spatial arrangement of visible GPS NAVSTAR satellites

Contagion A measure of the degree of clumping of land cover or vegetation types

Contrast (between patches) Differences in attributes of patches

Core habitat The central part of a habitat with very predictable (typical) conditions

Corridor A narrow strip of habitat surrounded by habitats of different types

Corridor patch A patch or a habitat that has the functions of a corridor

Covert Site in which three or more habitats meet

Cultural landscape A landscape profoundly changed by a long history of feedback between ecological processes and human activity, e.g. agriculture, forestry and pastoralism

Data dictionary A description of features and objects useful to store field survey data quickly in a GPS datalogger

Datalogger A handheld lightweight data entry computer, also used in GPS applications

Dehesa A belt of mosaic of pastures and scattered trees of central Spain. See also **Montado**

Differential correction Procedure to improve the accuracy of GPS data by combining data from a base station database of known coordinates and rover data collected at the same time. With this procedure the selective availability is removed

Discharge (water, nutrients) The rate of discharge of water or nutrients from a porous medium

Discontinuity The abrupt change of some characters of a system

Dispersion The capacity of individuals and populations to move to a new habitat or to new parts of a landscape

DOC Dissolved organic carbon

DoD Department of Defense. Operates the NAVS-TAR satellite for GPS

Ecodiversity The diversity of land cover type or forest type. May be used also in a cultural landscape to describe the diversity of land use and human culture integrated with ecological processes

Ecosphere Portions of the universe favourable for living organisms and in which all ecological processes are contained

Ecotone A transition site between different habitats, a tension zone between systems of different maturities and where energy exchange and material are highest

Ecotope The elementary unit of a landscape, homogeneous for a particular pattern or function

Edaphic (factor) Physical and chemical conditions of soil

Edge effect The presence of higher concentrations of organisms at the edges of a patch

Elevation mask The angle below which a GPS receiver does not track satellites. Rover receiver generally is set to 15°

Entropy A measure of landscape disorder or unavailable energy in a thermodynamic system

Ephemeral A phenomenon or an organism lasting for only a short time (few days or hours)

Evaporation The process of transformation of liquid into vapour

Evapotranspiration The loss of water by transformation into vapour by plant transpiration

Evenness The distribution of abundance between a collection of organisms or patches of a landscape

File A collection of related information stored in a computer, with a specific name

Florida Keys Coral islets or barrier islands off the southern coast of Florida

Fossorial behaviour The digging or burrowing behaviour of mammals

Fractal An object that has fractional dimensions and which at a changing scale of resolution shows self-similarity

Fragility An attribute of ecological systems: a system is fragile when under a perturbation regime a change in biological diversity occurs

Fragmentation A process by which forest cover is opened and isolated woodlots are created

Functional heterogeneity Heterogeneity in the spatial distribution of ecological entities (individual, populations, species, communities)

Functional patch A patch that has homogeneous characters for a particular function

Fymbos Shrub cover in South Africa (Cape Town region), similar in shape to Mediterranean maquis

Gamma diversity The diversity of species in different habitats along a geographical area

Gap phase The process that follows tree fall in forests and which produces clearings and regrowth by secondary succession

Geomorphic processes Events that modified the chemical substrate and physical appearance of facies, rocks and soils

GIS Geographical information systems

GPS Global positioning system, a satellite-based positioning system

Grain The resolution of an image or the minimum area perceived as distinct by an organism

Grid map A map in which data are stored in the form of grid cells

Guild A group of animals with similar characters associated to functions (foraging guild, breeding guilds etc.)

Habitat patch A patch selected by individuals of the same species

Hardwood forest The wood of angiosperm trees

HDOP Horizontal dilution of precision, attribute of NAVSTAR constellation

Hedgerow A strip of shrubs or trees planted in a rural landscape for signal properties or to protect crops from windstorms

Holartic (forest) Forest of Paleartic and Neartic zoogeographical regions (sin. northern hemisphere)

Holon Component of the horizontal structure of a hierarchical system

Home range The area in which a species normally lives

HRV High resolution visible scanner

Inbreeding Genetic exchange within related individuals

Incorporation The process by which a system reduces the effects of disturbance

Interrefuge corridors Strips of land that connect habitat patches considered as refuges for some species

Interior species Species living far from forest edges

Isotropic An object that is a rescaled copy of itself in all directions

Keystone species Species that shapes the habitat in which it lives and allows the presence of other species

Labels The description of an object represented in a map

Land unit The association of ecotopes. Synonymous with microchore

Landmarks Objects in a landscape used by organisms for orientation

Landscape patchiness A land mosaic composed of many patches

Layer A map component of a GIS system

Litter Vegetation material recently fallen on the ground and only partially decomposed

Local extinction The disappearance of a species from a patch

Local uniqueness The presence of unique characters linked to a particular site

Long-term ecological studies Studies planned in particular sites, regions and areas in order to track ecological processes over a long period of time

Macrochore A region composed of an aggregation of mesochores

Macroscale The level between meso- and megascale

Matrix The dominant component of a landscape mosaic

Megascale The upper level of scaling

Mesochore An aggregation of microchores

Mesolithic Archeological period from about 10 000 to 4000 years bp

Mesoscale An intermediate level between micro- and macroscale

Metapopulation Subpopulations that are connected by movement (immigration–emigration of individuals)

Microchore An aggregation of ecotopes

Microcosm A small world or community compared to a larger dimension or entity

Microscale The lower level of scaling

Montado The Portuguese part of the Spanish Dehesa. (See **Dehesa**)

MSS Multispectral scanner, a device mounted on LANDSAT satellites

Naturalness Attribute of land meaning intactness or integrity of ecosystems

NAVSTAR The official name of the GPS satellites. Acronym for Navigation Satellite Time And Ranging

NOAA National Oceanic and Atmospheric Administration (USA)

Nothofagus Genus of evergreen or deciduous of family Fagaceae

Nutrients (in the soil) Elements necessary for plant nutrition

Observation scale The spatiotemporal scale at which a process or a pattern is more easily observed (studied)

Outbreaks Organism demographic explosion, generally refers to pests (mice, insects, weeds)

Parish A religious division of a landscape common to all western Europe. One or more villages pertain to a parish

Pathogen Any microorganism that produces disease

PDOP Position dilution of precision, attribute of NAVSTAR constellation

Pedon The smallest unit or volume of a soil profile

Percolation The property of fluid to occupy a porous medium

Percolation threshold The value of 0.5928, calculated on large theoretical lattices, by which a fluid percolates, moving from one side of a matrix to the other

Perturbation A discrete event that modifies the status of a system without catastrophic consequences. Synonymous with Disturbance

Petrocalcic horizon A diagnostic subsurface of soil horizon characterized by induration with calcium carbonate

pH The negative logarithm ($\log_{10}$) of hydrogen-ion activity in solution

Phreatic water Synonymous with ground water, the water in the soil in the zone of saturation

Physiotope A land unit homogeneous for soil characters (see **Ecotope**)

Phytoplankton The plankton plants living free in water

Pixel Contraction of picture element, the smallest unit of information of a map or a raster image

Polypedon A group of contiguous pedons

Raster A representation by grid cells of an object in a computer memory

Resilience A process by which a system incorporates a disturbance by small changes in internal structure and function

Rover A mobile GPS receiver

SA Selective availability. A deliberate error introduced into GPS measurements by the DoD. SA can be completely removed by using the differential correction (see **GPS** snd **DoD**)

Savanna Tropical vegetation dominated by grasses and tall srubs with a different density of isolated trees

Seepage The slow movement of water in a porous medium (soil, litter)

Self-organized A property of a system to transform, conserve and transfer information

SEPM Spatially explicit population models

Shifting-mosaic steady state The condition in which a landscape changes in distribution of patches due to different causes, but in the end maintains the same character

Shrub-obligate species Species (e.g. birds) living exclusively in shrubs

Sink A population that becomes extinct without external immigration. May also refer to habitats

Source A population that has a positive balance between births and deaths. May also refer to habitats

Spacing The ability of an organism to react to its perception of the neighbouring environment

Spatial heterogeneity The variation across space of vegetation type or land cover

SPOT Satellite Pour l'Observation de la Terre, a French satellite

Stopover migrants Migratory birds that spend a short time in selected habitats along the migratory route to replenish energy

Structural patch A patch characterized by a recognizable pattern (for instance a soil type associated with a particular plant)

Supervised classification A method of remote sensing by which a satellite image is classified using training sets of classified patterns, land cover or vegetation

Synecology The study of the ecology of groups, populations and communities

TDOP Time dilution of precision, attribute of NAVSTAR constellation

Temporal heterogeneity The variation across time of a vegetation type or land cover

Terracette A levelled piece of land on a steep slope protected downwards by a stone or mound wall

The Everglades The Florida marshes and wet forests subject to periodic freshwater flooding

Thematic map A geographical representation of land use or vegetation cover, or other natural or socioeconomic phenomenon

TM Thematic Mapper, a sensor place on LAND-SAT satellite

Topology Pattern of linkage between geographical elements

Total human ecosystem Conceptualization of modifications and effect of human life on the earth

Trampling Soil compaction due to animal passage

Traps Habitats that attract species because of their favourable conditions but in which some functions such as breeding are not allowed or are suddenly interrupted by human disturbance or predation

Tree-fall gap Opening in forest cover due to individual tree fall

UTM Universal transverse Mercatore projection

VDOP Vertical dilution of precision, attribute of NAVSTAR constellation

Vector An object that has magnitude and direction

REFERENCES

Allaby, M. (1985) *The Oxford dictionary of natural history*. Oxford University Press, Oxford.

Bates, R.L. and Jackson, J.A. (1987) *Glossary of geology*, 3rd edn. American Geological Institute, Alexandria, VA.

Lincoln, R.J., Boxshall, G.A., Clark, P.F. (1982) *A dictionary of ecology, evolution and systematics*. Cambridge University Press, Cambridge.

Index

Note: page numbers in **bold** refer to figures, those in *italics* refer to tables.